PLEASE STAMP DATE DUE, BOTH

DATE DUE | DATE DUE

MAR

Optical System Design

Optical System Design

RUDOLF KINGSLAKE

Institute of Optics
University of Rochester
Rochester, New York

ACADEMIC PRESS, INC.
(Harcourt Brace Jovanovich, Publishers)
Orlando   San Diego   New York   London
Toronto   Montreal   Sydney   Tokyo

COPYRIGHT © 1983, BY ACADEMIC PRESS, INC.
ALL RIGHTS RESERVED.
NO PART OF THIS PUBLICATION MAY BE REPRODUCED OR
TRANSMITTED IN ANY FORM OR BY ANY MEANS, ELECTRONIC
OR MECHANICAL, INCLUDING PHOTOCOPY, RECORDING, OR ANY
INFORMATION STORAGE AND RETRIEVAL SYSTEM, WITHOUT
PERMISSION IN WRITING FROM THE PUBLISHER.

ACADEMIC PRESS, INC.
Orlando, Florida 32887

United Kingdom Edition published by
ACADEMIC PRESS, INC. (LONDON) LTD.
24/28 Oval Road, London NW1 7DX

Library of Congress Cataloging in Publication Data

Kingslake, Rudolf.
  Optical system design.

  Bibliography: p.
  Includes index.
  1. Optics. 2. Optical instruments. I. Title.
QC371.K52 1983   535'.33   83-9997
ISBN 0-12-408660-8

PRINTED IN THE UNITED STATES OF AMERICA

84 85 86 87   9 8 7 6 5 4

Contents

Preface   ix

1. Optical Systems

    I. DESIGN AND PRODUCTION   1
    II. OPTICAL MATERIALS   2
    III. LENS MANUFACTURE   3

2. Light and Images

    I. THE NATURE OF LIGHT   7
    II. THE LAW OF REFRACTION   10
    III. A PERFECT OPTICAL SYSTEM   13
    IV. LENS ABERRATIONS   21
    V. FIBER OPTICS   22

3. Ray-Tracing Procedures

    I. TYPES OF RAYS   26
    II. GRAPHICAL RAY TRACING   27
    III. MERIDIONAL RAY TRACING   28
    IV. PARAXIAL RAYS   30
    V. CURVED MIRRORS   36
    VI. MAGNIFICATION AND THE LAGRANGE THEOREM   36
    VII. THE FRESNEL LENS   38

4. The Gaussian Theory of Lenses

    I. INTRODUCTION   40
    II. THE FOUR CARDINAL POINTS   40
    III. CONJUGATE DISTANCE RELATIONSHIPS   47
    IV. A SINGLE LENS   51
    V. LONGITUDINAL MAGNIFICATION   56
    VI. THE SCHEIMPFLUG CONDITION   58
    VII. FOCOMETRY   59
    VIII. AUTOFOCUS MECHANISMS   62

5. Multilens Systems

    I. GRAPHICAL CONSTRUCTION OF AN IMAGE   66
    II. RAY TRACING THROUGH A SYSTEM OF SEPARATED THIN LENSES   67
    III. TWO THIN LENSES WITH A REAL OBJECT AND REAL IMAGE   69
    IV. THE MATRIX APPROACH TO PARAXIAL RAYS   77
    V. CYLINDRICAL LENSES   80

6. Oblique Beams

    I. MERIDIONAL RAYS   84
    II. THE IRIS AND PUPILS OF A LENS   86
    III. PARAXIAL TRACING OF AN OBLIQUE BEAM   89

7. The Photometry of Optical Systems

    I. INTRODUCTION   92
    II. PHOTOMETRIC DEFINITIONS   93
    III. PHOTOMETRIC PROPERTIES OF PLANE SURFACES   98
    IV. PHOTOMETRIC MEASURING INSTRUMENTS   110
    V. THE FLUX EMITTED BY A PLANE SOURCE   117
    VI. ILLUMINANCE DUE TO A CIRCULAR SOURCE   120
    VII. ILLUMINATION IN AN OPTICAL IMAGE   124

8. Projection Systems

    I. A SELF-LUMINOUS OBJECT   131
    II. AN ILLUMINATED OBJECT   131

    III. PROJECTION SCREENS   138
    IV. STEREOSCOPIC PROJECTION   140
    V. CONTOUR PROJECTORS   141

9. Plane Mirrors and Prisms

    I. RIGHT- AND LEFT-HANDED IMAGES   144
    II. ROTATING MIRRORS   148
    III. MULTIPLE MIRROR SYSTEMS   153
    IV. REFLECTING PRISMS   155
    V. IMAGE ROTATORS   161
    VI. PRISMATIC IMAGE ERECTORS   164

10. The Eye as an Optical Instrument

    I. DIMENSIONS   170
    II. THE PROPERTIES OF VISION   172
    III. SPECTACLE LENSES   176
    IV. STEREOSCOPIC VISION   180

11. Magnifying Instruments

    I. THE SIMPLE MAGNIFIER   182
    II. THE COMPOUND MICROSCOPE   187
    III. ABBE THEORY OF MICROSCOPE VISION   196
    IV. MICROSCOPY OF TRANSPARENT OBJECTS   199

12. The Telescope

    I. FUNDAMENTAL PROPERTIES   203
    II. EYEPIECES   208
    III. ERECTING TELESCOPES   210
    IV. OTHER TYPES OF TELESCOPES   221
    V. ASTRONOMICAL TELESCOPES   227

13. Surveying Instruments

    I. CLASSES OF SURVEYING INSTRUMENTS   230
    II. RANGEFINDERS   236
    III. AN AXICON   243

14. Mirror Imaging Systems

  I. SINGLE-MIRROR SYSTEMS   245
 II. TWO-MIRROR SYSTEMS   253
III. COMA CORRECTION   256
 IV. OBSTRUCTION   259

15. Photographic Optics

   I. PERSPECTIVE EFFECTS IN PHOTOGRAPHY   262
  II. FOCUSING ON A NEAR OBJECT   263
 III. DEPTH OF FOCUS AND DEPTH OF FIELD   265
  IV. THE THEORY OF TILTED PLANES   269
   V. SPECIAL PURPOSE LENSES   272
  VI. TYPES OF ZOOM LENSES   277
 VII. SPECIFYING A PHOTOGRAPHIC OBJECTIVE   280
VIII. PANORAMIC OR SLIT CAMERAS   281
  IX. MOTION-PICTURE SYSTEMS   286

16. Spectroscopic Apparatus

  I. DISPERSING PRISMS   294
 II. DIFFRACTION GRATINGS   306

Index   315

Preface

It often happens in industrial and government laboratories that an electrical or mechanical engineer is required to design an optical system to perform some particular function. He then needs a basic knowledge of optics, especially geometrical, and a clear understanding of the flow of light through an optical system. The purpose of this book is to provide this basic information, enabling a nonoptical engineer to tackle his unfamiliar problem.

Several standard types of optical instrument are described in detail because they are so often used alone or in combination to form more complicated systems. Some degree of familiarity with physical optics, diffraction, and coherence is also required because of the importance of phenomena resulting from the finite wavelength of light. These matters receive consideration wherever they arise in the discussion of a problem.

This book represents an expansion of undergraduate and summer-school courses in geometrical optics and system design given by the author in the Institute of Optics at the University of Rochester. No attempt has been made to include lens design; that has been dealt with in the author's "Lens Design Fundamentals" (Academic Press, 1978).

It is clearly impossible to explain every conceivable situation in a single volume, and the reader will need to refer frequently to more specialized sources. Much of this additional information will be found in the nine volumes published since 1965 by Academic Press, and the future volumes of "Applied Optics and Optical Engineering." Other useful sources are the regular issues of the journals *Optical Engineering*, published by SPIE, and *Applied Optics*, published by the Optical Society of America. Excellent further references are Warren Smith's book "Modern Optical Engineering" (McGraw Hill, 1966), and "Military Standardization Handbook 141: Optical Design" (Defense Supply Agency, Washington, 1963). For work in the infrared, "The Infrared Handbook," edited by W. L. Wolfe and G. J. Zissis (ERIM, Ann Arbor, 1979) will be found useful.

CHAPTER 1

Optical Systems

I. DESIGN AND PRODUCTION

The production of an optical system, whether a unique instrument or a mass-production item, necessarily follows three well-defined stages: (a) system layout, (b) lens design, and (c) optical engineering.

In the first of these stages, the structure of the system is defined so that it will meet the customer's requirements and perform its desired function. The *system designer* must be highly skilled in optics, knowledgeable as to the availability of components, and aggressive in keeping costs to a reasonable level. He is ultimately responsible for the entire project.

When the structure of the system has been established, the *lens designer* takes over and fills in the details of each lens or other optical component that will be required. Finally, the *optical engineer* is called in to design the mechanical parts and to decide how everything is to be made, either in-house or purchased from an outside supplier. He must also set up a time schedule and plan the various test procedures that will be used before the release of the instrument to the customer. In a small establishment these three functions may be combined in a single individual, whereas in a large company each function may be the responsibility of a whole group. Nevertheless, all persons involved must work together to ensure that the job is completed on time, that the costs are kept to a minimum, and that the system will operate reliably and satisfactorily.

Besides these people, there will be the usual departments for estimating, drafting, factory planning, cost accounting, etc., which apply to everything that is undertaken in the establishment, but the three functions listed above apply particularly to the design of an optical system. In this book only the first of these functions will be considered.[1]

---

[1] The functions of the lens designer are discussed in "Lens Design Fundamentals" by Kingslake (Academic Press, 1978). Optical engineering is covered broadly in the series of volumes entitled "Applied Optics and Optical Engineering," edited by Kingslake (vol. 1–6) and by Shannon–Wyant (vol. 7–9) (Academic Press, 1965–1983).

I. OPTICAL SYSTEMS

An *optical system* is an assembly of components working together to produce a desired result. This may be as simple as forming an image having a specified brightness and size at a given location, or it may be a complex system involving some or all of the following features:

(1) light sources, lamps, lasers, LEDs;
(2) radiation detectors, photocells, thermocouples, photographic emulsions;
(3) lenses and curved mirrors for image formation;
(4) plane mirrors and prisms to deflect light or rotate an image;
(5) projection and display devices, screens, light valves, CRTs, liquid crystals;
(6) thin films;
(7) polarizers and retardation plates;
(8) chromatic and achromatic filters;
(9) beam splitters;
(10) image scanners, mirrors, polygons, acoustooptics;
(11) image tubes, IR, UV, x ray, light amplifiers;
(12) television tubes, Vidicons, Orthicons;
(13) charge-coupled devices;
(14) Pockels and Kerr cells;
(15) gratings and prisms to form a spectrum;
(16) fibers, face plates, waveguides;
(17) integrated optics;
(18) image processing, pattern recognition, spatial filtering;
(19) stereoscopy, photogrammetry;
(20) interferometers, holograms;
(21) motion-picture equipment, sound recording and reproduction.

II. OPTICAL MATERIALS

Most lenses are made of optical glass, several hundred types of which are produced regularly by many manufacturers around the world; all necessary data about these can be found in the makers' catalogs. Millions of lenses are made every year by injection molding of plastic materials, but unfortunately, the range of available types of plastic is severely limited and the refractive indices are generally low. The high temperature-coefficients of expansion and refractive index can also present problems. Occasionally, crystalline materials such as quartz, sapphire, or lithium fluoride are used, either because of their hardness or their transparency in the ultraviolet. Such materials as silicon and germanium are used in the infrared because of their transparency in that region, but these materials are likely to be opaque in the visible region, and lenses made from them are difficult to test and adjust. It has often been proposed that a lens be made of a thin shell, spherical or aspheric, filled with liquid, but thermal effects and the possibility of leakage render this rather impractical. All in all, glass provides the greatest range of useful materials in the wavelength range from 0.36 to 2.5 μm.

Recently, experiments have been started on the making of lenses with materials having either a longitudinal or a transverse gradient of refractive index. Both types have their uses, and work has been undertaken on the routine fabrication of suitable materials. If these become practical, they will replace aspheric surfaces in lenses. Gradient-index (GRIN) rods are already common, and gradient optical fibers are being used as communication channels between cities that are many miles apart.

III. LENS MANUFACTURE[2-4]

To make a lens having spherical surfaces, a suitable piece of material is molded into a circular form, and the surface radii are generated by the use of a diamond-studded tool of ring shape, as indicated in Fig. 1.1a. The lens surfaces are then lapped with loose emery on an iron tool shaped to the desired curve, the tool being the reverse of the desired lens surface. The tool is oscillated back and forth while the lens is rotated about a vertical axis, with the most irregular motion producing the best approach to a perfect spherical surface (Fig. 1.1b). After lapping is complete, the surfaces are polished with a pitch-covered tool, using cerium oxide or rouge as a polishing agent.

The difference between lapping and polishing is that in the former operation the abrasive grains are loose and free to roll over and chip the

[2] F. Twyman, "Prism and Lens Making." Hilger and Watts, London, 1952.
[3] D. F. Horne, "Optical Production Technology." Crane Russak, New York, 1972.
[4] A. S. DeVany, "Master Optical Techniques." Wiley, New York, 1981.

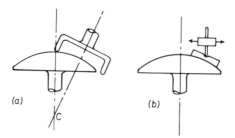

FIG. 1.1. (a) Curve generation. (b) Lapping and polishing.

glass, whereas in the latter the grains are stuck into the surface of the pitch and tend to scrape or "sandpaper" the glass surface. The polishing process can be speeded up by the application of high pressure and rapid relative movement of the tool and the glass.

A. Centering and Edging

After grinding and polishing is complete, a lens is edged to the required diameter by rotating it about an axis containing the centers of curvature of the two surfaces. The completed lens elements are then mounted in a metal cell so that all the elements share a common axis (Fig. 1.2a). Failure to do this will result in very poor image quality, which can be caused by a decentered element or even by a slightly tilted surface (Fig. 1.2b). Centering and mounting lenses is the most difficult part of the manufacturing process.

B. Aspheric Surfaces

Any attempt to generate an aspheric surface of revolution by normal methods will result in a spherical surface unless some means are found to restrict the relative motion of the units. Many companies are developing ways to manufacture aspheric surfaces, but the problem is complex and such lenses are liable to be very expensive.[5] Not the least of these problems is the development of a test procedure to determine

[5] R. R. Shannon, "Aspheric surfaces," *in* "Applied Optics and Optical Engineering" (R. R. Shannon and J. C. Wyant, eds.), Vol. 8, p. 55. Academic Press, New York, 1980.

III. LENS MANUFACTURE

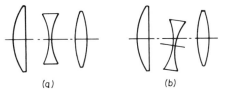

FIG. 1.2. (a) A centered lens. (b) A lens with one element tilted and displaced.

whether the surface generated is the correct one or if it is so far from the desired shape as to introduce intolerable aberrations into the image. An aspheric surface has its own unique axis which must be mounted so that it lies in the axis formed by the centers of curvature of all the other surfaces in the system. Plastic lenses can be molded with an aspheric surface as easily as with a spherical surface, but the mold must be correctly formed and this is as difficult as making an aspheric lens. However, once the mold has been made, tested, and found to be correct, thousands of lenses can be produced with very little difficulty. Rather than making the whole lens out of plastic materials it has been proposed to deposit a thin layer of plastic material on a polished glass substrate and mold the layer to the desired aspheric shape.

A cylindrical surface can be ground and polished to a circular section by ordinary methods, provided the grinding and polishing tools are prevented from rotating. A cylindrical lens can be made very long and cut into sections after completion, with each section then being edge-ground to the required diameter.

C. Aligning an Optical System

An important step in the manufacture of an optical system is the alignment of a series of lenses and mirrors in the housing. The usual procedure is to employ a small milliwatt gas laser, which emits a narrow parallel beam, and to make a small ink dot in the middle of each lens in the system. By sending the laser beam along the system axis, it is readily seen whether the beam passes through all the ink dots. Finer tuning may be required later if the image shows signs of coma on the axis, which obviously should not be there.

The laser procedure is particularly useful if plane mirrors are used within a system to deflect a beam. For instance, in the system shown in

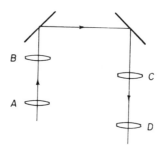

Fig. 1.3. Aligning an optical system.

Fig. 1.3, which contains several lenses and two mirrors at 45°, the laser beam is sent along the axis of lenses A and B and then at the two mirrors. The beam can be brought through the middle of lens C by adjusting either mirror, but only when both mirrors are correctly oriented will the beam pass through the middles of both lens C and lens D.

CHAPTER 2

Light and Images

I. THE NATURE OF LIGHT

Light is a wave motion traveling through space or other transparent medium that is able to affect the eye and certain artificial detectors to form an image. The principal sources of light are the sun, sunlight reflected from surrounding objects, tungsten lamps, carbon arcs, lasers, gas discharge tubes, and fluorescent and phosphorescent materials excited by electrons or UV radiation. The principal detectors of radiation are eyes, thermal detectors, photocells, and photographic emulsions. Some detectors merely respond to radiation, whereas others retain an image for further study.

Light waves have a frequency lying between 7.5×10^{14} vibrations/sec for violet light and 4×10^{14} vibrations/sec for deep red light. The wavelength, or distance from crest to crest, is found by dividing the velocity of light (3×10^{10} cm/sec in empty space) by the frequency; thus in air the wavelength of violet light is about 0.4 μm (micrometer) and in red light it is about 0.75 μm. In glass or other transparent media the wavelength and the velocity are reduced by a factor known as the *refractive index* of the material. Refractive indices range from as low as 1.33 for water to as high as 4.0 for germanium in the infrared. Few transparent materials have refractive indices exceeding 2.0, although diamond has an index of about 2.5. Ordinary window glass has a refractive index of about 1.52.

A measure of light waves that is particularly useful to infrared spectroscopists is the *wave number,* or number of waves in a centimeter. This ranges from 25,000 for violet light to about 13,000 for red light. In the far infrared region the wave numbers become smaller and more manageable; for instance, at a wavelength of 0.1 mm the wave number is 100, and at 1 mm it has dropped to 10.

A. Polarized Light

Light is a transverse wave motion, and it is possible by suitable means to cause the vibrations to lie in one direction only. Such *plane-polarized* light has many interesting properties. Ordinary light can be regarded as a mixture of vibrations lying in all possible orientations.

Since 1808 it has been known that light becomes partially polarized when it is obliquely reflected from glass or other nonmetallic material. It is also partially polarized by internal strain in glass and by some transparent crystals, and this latter property is employed in Polaroid material, which is used extensively to polarize light in instruments. Some crystals, such as calcite, are strongly *birefringent;* they have entirely different refractive indices for light vibrating parallel and perpendicular to the optic axis of the crystal, and prisms made of calcite have been used for 150 years to polarize a beam of light.

Two polarizing prisms, or Polaroid sheets, can be used in succession to vary the intensity of a beam of light in a quantitative manner. If the two polarizers are set with their vibration directions parallel, the first polarizer reduces the intensity of the light to half, because half is lost by reflection or absorption, but the second polarizer has no further effect on the intensity. However, when the second polarizer, which is often called the analyzer, is rotated about the axis of the beam, the intensity of the transmitted light varies as the $\cos^2$ of the angle between the two polarizing directions and becomes completely dark when the polarizing directions are at right angles. This property is quite quantitatively accurate and can be used for making photometric measurements.

If three polarizers are used in succession, the first two determine the maximum transmittance of the assembly, and the third serves to reduce this maximum to zero when the second and third polarizers are crossed.

B. Rays and Waves

Although light is known to be a wave motion, it is often more convenient to consider only the paths along which light travels. These paths are known as *rays,* and in a homogeneous medium they are straight lines. However, the ray concept is basically a mathematical fiction, and the highly convenient ray treatment of imaging systems necessarily ignores all wave-related phenomena such as polarization, diffraction, interference, and scattering. We can readily determine the location and brightness of an image by ray methods, but if we wish to study the fine structure and distribution of light within an image, we must resort to a wave treatment of the situation, and this is generally a far more difficult operation. The study of ray phenomena is known as *geometrical optics,* and the study of light waves constitutes *physical optics.*

C. Real and Virtual Images

The image of some given object formed by an optical system can be *real,* i.e., formed outside the system where it can be allowed to fall on a receiving surface such as a screen or a piece of film, or it may be *virtual,* i.e., formed inside the system where it can be observed by looking into the end of the system (Fig. 2.1) Examples of the former are cameras and projectors, and examples of the latter are eyeglasses, telescopes, and microscopes.

Similarly, a real object lies outside the system in the usual way, but an object can also be virtual if it is projected into the system by some external means. Of course, in any system containing several refracting surfaces, the image formed by one surface becomes the object for the next, but the terms "real" and "virtual" are generally restricted to the original object and final image of a system. Actually, only real images are of any value, and the virtual image formed in a microscope becomes a real image on the retina of the observer's eye.

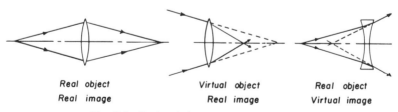

Real object   Virtual object   Real object
Real image   Real image   Virtual image

FIG. 2.1. Real and virtual objects and images.

D. Object Space and Image Space

Considered naively, the *object space* and the *image space* of a lens system are those spaces containing the object and image, respectively. This is fine if the object and image are real, but if they are virtual, it becomes somewhat confusing. We may regard the object space, no matter where the object itself may be located, as the space containing the incident rays, and likewise the image space as the space containing the emerging rays. Alternatively, we can regard the object space as the space containing the object, but if the object is virtual, we must postulate that the object and image spaces overlap each other to infinity in both directions—a concept that some people find hard to accept. Of course, in a spherical mirror, for example, the entering and reflected rays necessarily lie on the same side of the mirror, and therefore the object and image spaces must be considered to overlap, and the confusion cannot be avoided.

II. THE LAW OF REFRACTION

Let us consider a plane wave front, in a parallel beam of light, that is incident upon a plane refracting surface, as indicated in Fig. 2.2. At the instant of time when A, the bottom of the wave, touches the surface, then B, the upper end of the wave, will be distant from the surface by

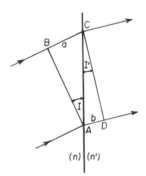

FIG. 2.2. The law of refraction.

an amount a. While B moves to C in the left-hand medium, the lower end of the wave moves from A to D in the right-hand medium, a distance b. Because the velocity of light is different on the two sides of the refracting surface, a will not be equal to b; in fact, a/b is equal to the ratio of the two refractive indices n'/n.

Turning now to angles, the angle of incidence I between the incident ray and the normal to the surface at the point of incidence, and the corresponding angle of refraction I', are related by

$$\frac{\sin I}{\sin I'} = \frac{BC}{AC} \bigg/ \frac{AD}{AC} = \frac{BC}{AD} = \frac{a}{b} = \frac{n'}{n},$$

hence

$$n \sin I = n' \sin I'.$$

This is the well-known *law of refraction,* first stated in this form by Descartes in 1637, although a geometrical construction for the direction of the refracted ray had been given by W. Snell in 1621; it is often referred to as Snell's law. This law underlies all considerations of optical systems based on ray paths.

As already mentioned, a ray is a purely mathematical concept, and it represents the path along which a light wave will travel. The rays are everywhere perpendicular to the wave fronts, and because of the symmetry of the law of refraction, we can assume that light can travel in either direction along a ray.

A. Total Internal Reflection

It is clear from Fig. 2.3 that if a ray in a denser medium of refractive index n is refracted at a surface into a medium of lower refractive index n', the angle of refraction I' will be greater than the angle of incidence I, and when I reaches the limiting value of $\sin I = n/n'$, the angle of refraction I' will be just equal to 90° and the refracted ray will lie along the surface. This particular value of the angle of incidence in the denser medium is called the *critical angle* ϕ. At angles of incidence smaller than ϕ the ray is refracted out in the ordinary way, but if I in the denser medium is greater than ϕ, there will be no refracted ray, and the whole of the light will be reflected back into the denser medium, as indicated

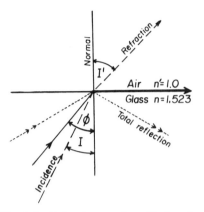

FIG. 2.3. Refraction and the critical angle.

by the dashed ray in Fig. 2.3. This subject is discussed much more fully in Chapter VII, Section IIIA.

B. Interpolation of Refractive Indices

Optical glass catalogs give refractive index data for a large number of wavelengths, from 0.365 μm in the near ultraviolet to 1.014 μm in the infrared. However, it is sometimes necessary to interpolate between the given wavelengths, and for this purpose a formula connecting refractive index with wavelength is needed. In the catalog of the Schott Optical Glass Company, a six-term formula is used of the form

$$n^2 = A_0 + A_1\lambda^2 + A_2/\lambda^2 + A_3/\lambda^4 + A_4/\lambda^6 + A_5/\lambda^8. \quad (1)$$

All six coefficients A_0 to A_5 are given explicitly to eight significant figures for each type of glass. The terms with λ in the numerator become large in the infrared, and those with λ in the denominator become large in the ultraviolet. The index of most glasses rises rapidly in the UV but drops only slowly in the IR, which explains why in Eq. (1) four UV terms are used with only one in the IR.

Some other interpolation equations have been proposed, but to apply them it is necessary to solve for the various coefficients by use of an appropriate number of stated refractive indices at known wavelengths.

III. A PERFECT OPTICAL SYSTEM

A perfect optical system is one in which all the light rays from an object point on one side of the system pass through a single image point on the other side of the system. The emerging wave front is then spherical, and the optical paths from object to image are all equal. The *optical path* is the sum of the products nD along each ray, where n is the refractive index of any portion of the ray path and D the length of that portion.

In the case of a nonperfect system, the optical paths will not all be equal nor will all the rays pass through a single image point, and the emerging wave front will not be spherical. We can express the quality of such an image by the scatter of the emerging rays or by stating the maximum difference between the various possible optical paths from object point to image point. Lord Rayleigh found that if the optical path difference (often referred to as OPD) is everywhere less than a quarter of the wavelength of the light, the image will not be noticeably imperfect; this is commonly known as the *Rayleigh limit of permissible imperfection* in an image.

A. THE AIRY DISK

It is shown in books on physical optics that the image formed by a perfect optical system at the center of curvature of the emerging spherical wave front is not a point, as geometrical optics suggests, but instead it is a tiny circular spot of light surrounded by a series of very faint rings of light having rapidly diminishing intensities. This image structure is called an *Airy disk* after the man who first worked out its properties in 1834. Theoretically, the succession of concentric rings goes on forever, but after two or three rings the intensity becomes so low that they are no longer visible.

A graph of the light intensity across the Airy disk is shown in Fig. 2.4. It is seen that the central spot is enormously brighter than even the first ring, the brightness of which is only about $\frac{1}{60}$ of the brightness of the central spot, while the other rings are even fainter still. It is therefore virtually impossible to photograph an Airy disk, for if the exposure is adjusted to record the central spot, the rings will not be recorded, and if the exposure is set to photograph the rings, the central

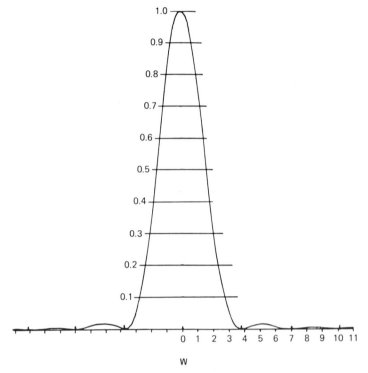

FIG. 2.4. Section of the Airy disk.

spot will be so broadened by halation that it will swallow up the rings entirely.

The equation representing the brightness of the Airy disk system at any radial distance w from the center is

$$I = \left[\frac{2J_1(w)}{w}\right]^2,$$

with the normalizing factor 2 being introduced to make the central intensity equal to 1.0. The symbol J_1 refers to the first-order Bessel function, tables of which are readily available. The successive dark rings fall at the zeros of the Bessel function, which are $w = 3.832$, 7.016, 10.173, 13.324, 16.471, etc. A series of values of the Bessel function and of the intensity of the Airy disk are given in Table I. The most remarkable feature of the Airy disk is the large amount of light

III. A PERFECT OPTICAL SYSTEM

TABLE I

VALUES OF THE BESSEL FUNCTION AND OF THE INTENSITY OF THE AIRY DISK

w	$J_1(w)$	Intensity	w	$J_1(w)$	Intensity
0.0	0.0	1.0	6.5	−0.154	0.0022
0.5	0.242	0.939	7.0	−0.005	0.0000
1.0	0.440	0.775	7.5	0.135	0.0013
1.5	0.558	0.553	8.0	0.235	0.0034
2.0	0.577	0.333	8.5	0.273	0.0041
2.5	0.497	0.158	9.0	0.245	0.0030
3.0	0.339	0.0511	9.5	0.161	0.0012
3.5	0.137	0.0062	10.0	0.044	0.0000
4.0	−0.066	0.0011	10.5	−0.079	0.0002
4.5	−0.231	0.0106	11.0	−0.177	0.0010
5.0	−0.328	0.0172	11.5	−0.228	0.0016
5.5	−0.341	0.0154	12.0	−0.223	0.0014
6.0	−0.277	0.0085	12.5	−0.165	0.0007

energy in the outer rings. The Table II lists the radii of the successive rings and their relative brightness, and also the amount of energy remaining outside each dark ring. The actual physical value of the radius of any particular ring can be found from the formula

$$\text{radius of ring} = w(\lambda l'/\pi D) = (w/\pi)\lambda(F\text{-number}),$$

where D is the diameter of the lens aperture, l' the distance of the

TABLE II

VALUES OF VARIOUS PARAMETERS RELATED TO THE AIRY DISK

Ring	Radius of ring w	Relative intensity	Amount of light in ring (%)	Light remaining outside the ring (%)
Center	0.0	1.0	83.9	—
First dark	3.832	—	—	16.1
First bright	5.136	0.0175	7.1	—
Second dark	7.016	—	—	9.0
Second bright	8.417	0.0042	2.8	—
Third dark	10.173	—	—	6.2
Third bright	11.620	0.0016	1.5	—
Fourth dark	13.324	—	—	4.7

image from the lens, and λ the wavelength of the light. Thus, for example, the radius of the first dark ring for a perfect $f/4.5$ lens in sodium-D light with a distant object is found to be 0.0032 mm, or 3.2 μm.

The effective diameter of the central spot is usually stated to be about 70% of the diameter of the first dark ring; thus in the visible region it can be assumed to be about equal to the F-number of the lens expressed in micrometers. This amounts to a diameter of 4.5 μm in the example given. In the infrared, where the wavelength is so much greater, the effective diameter of the central spot becomes quite large, e.g., 38 μm for an $f/4.5$ lens at 10 μm in the IR. This explains why optical elements intended for use in the infrared "window" at 10 μm can have sizable surface errors that would be quite inadmissable in the visible region.

The special case of a central obstruction leading to an annular lens aperture is discussed in Chapter 14. This situation often arises in a mirror system such as a Cassegrain telescope.

B. A Diffraction-Limited Lens

A lens in which the aberrations are so small that the image of a point is no larger than the Airy disk is often referred to as being *diffraction limited*. Very few real lenses can make this claim over their whole field, but many are diffraction limited on the lens axis. Since aberrations are scaled in proportion to the focal length, whereas diffraction effects depend only on the relative aperture (F-number) of the lens, many lenses are diffraction limited when the focal length is short but not when they are made to a longer focal length. Microscope objectives are invariably diffraction limited but, again, only because their focal length is quite small.

C. Resolving Power of a Perfect Lens

Lord Rayleigh considered that two equally bright stars could be just resolved by a perfect telescope if the image of one star fell on the first dark ring of the Airy disk of the other star. Thus, he considered that the least resolvable separation (LRS) of the two star images in the focal plane of the telescope is given by

$$\text{LRS}' = \frac{3.832}{\pi} \lambda \left(\frac{l'}{D}\right) = 1.22\lambda\left(\frac{l'}{D}\right),$$

where D is the diameter of the lens aperture and l' the image distance from the lens. Actually, this is somewhat pessimistic and we can safely ignore the factor of 1.22. Expressed in terms of U', which is the slope angle of the steepest ray that can pass from the lens to the axial image point, we see that $\sin U' = \frac{1}{2}D/l'$, hence

$$\text{LRS}' = \tfrac{1}{2}\lambda/\sin U'.$$

Because the magnification is $m = (n \sin U)/(n' \sin U')$, where $n' = 1$ for air, we see that the least resolvable separation in the object is given by

$$\text{LRS} = \text{LRS}'/m = \tfrac{1}{2}\lambda/n \sin U.$$

In microscopy the product $n \sin U$ is called the *numerical aperture* in the object space, and correspondingly we may regard the numerical aperture in the image space as being equal to $n' \sin U'$. In this notation, the least resolvable separation in either the object space or the image space is given by half the wavelength divided by the numerical aperture in that space.

D. Modulation Transfer Function

As its name implies, the modulation transfer function (MTF) of a lens is a measure of its ability to form a faithful image of a given object. For simplicity we may assume that the object consists of a series of bars of light and darkness at progressively closer spacing, as indicated in (a) of Fig. 2.5, with the cross section of intensity being a sine wave. Since the image of a sine wave light distribution is always a sine wave, no matter how bad the aberrations may be, the image will have a sine-wave intensity distribution, as indicated in (b) of Fig. 2.5. The image of the bars is scanned by a narrow slit placed parallel to the bars, with a sensitive photocell behind the slit to measure and plot the light distribution in the image, with the output from the photocell resembling (c)

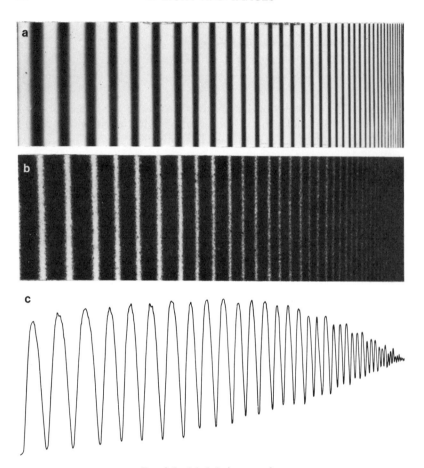

FIG. 2.5. Modulation transfer.

of Fig. 2.5. When the bars are coarse and widely spaced, the lens has no difficulty in accurately reproducing them, but as the bars get closer together, diffraction and aberrations in the lens cause some light to stray from the bright bars into the dark spaces between them, with the result that the light bars get dimmer and the dark spaces get brighter, until eventually there is nothing to distinguish light from darkness and all resolution is lost.

From the photocell response it is possible to plot a graph connecting image contrast with line frequency, where *contrast* is the intensity ratio

$$\text{contrast} = \frac{\text{bright minus dark}}{\text{bright plus dark}}.$$

It is easy to see that if the dark lines have zero intensity, the contrast is 1.0, and if the bright and dark lines are equally intense, the contrast is zero. The abscissa of our plot is the spatial frequency of the bars expressed as so many lines per millimeter in the image. Such a plot is shown in Fig. 2.6 for several points in the field of a photographic lens. It should be noted that, for a given lens, the plot of MTF differs by wavelength, by field obliquity, by the orientation of the bars, and from point to point along the lens axis. The curves of MTF can be plotted experimentally if suitable measuring equipment is available, or they can be computed if the lens structure is known. Lens designers are resorting more and more to calculated MTF curves as an indication of the anticipated performance of a lens while it is still in the design stage.

An incidental advantage of using MTF as a quality criterion is the ability to *cascade* the MTF curves of a lens and film (or other receiving surface) by multiplying together the MTF values, ordinate by ordinate, of lens and film at successive spatial frequencies. Indeed, in a typical photographic system, we can multiply frequency by frequency the MTF curves of the camera lens, the negative film, the printer lens, the positive material, and the observer's eye, giving us the net effect of the entire system from the original object to the perceived image. It must be particularly noted that the MTF curves of two successive lenses cannot be cascaded unless there is a diffusing layer between them. This is because the aberrations of one lens can be used to offset opposite aberrations in the other lens, so that two lenses that are separately

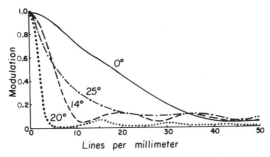

FIG. 2.6. Typical MTF curves for a photographic lens.

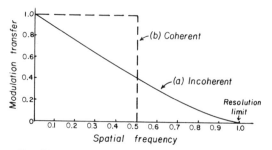

Fig. 2.7. MTF curves for a diffraction-limited lens.

imperfect can add up to a perfect result. If there is an aerial image between the two lenses, the aberrations may be positive or negative, whereas if there is a diffusing layer between them, then all aberrations must be regarded as positive and no cancellation can occur.

It should be noted that if the sine-wave bar chart is illuminated coherently, then everything is changed. As Thompson[1] pointed out, there is no longer any meaning to the ordinary intensity transfer function, or MTF curve, but we can plot a graph of the apparent, or amplitude, transfer function, the two graphs being shown together in Fig. 2.7. For any degree of partial coherence the curve lies between these two limiting situations.

Thus if the illumination is perfectly coherent, the resolution drops to half but the contrast at low frequencies is greatly improved. For many years microscopists have made use of this fact when trying to raise the contrast of a transparent object. By merely closing down the aperture of the substage condenser, the degree of partial coherence in the illumination is increased, and the contrast is improved but at the expense of resolution.

A better representation of the effect of coherence is to plot the gradient of an edge separating light from darkness.[2] Under incoherent illumination the edge gradient for a perfect lens has an S shape, sloping gradually from darkness to light. With coherent illumination, however, the slope is much steeper, with the formation of a series of rapidly disappearing fringes parallel to the edge on the light side of it. This phenomenon is known as *ringing*.

[1] B. J. Thompson, Image formation with partially coherent light, *in* "Progress in Optics" (E. Wolf, ed.), Vol. 7, p. 212. North Holland, Amsterdam, 1969.
[2] *Ibid.*, p. 216.

IV. LENS ABERRATIONS

There are seven well-established classes of aberrations, each of which contains a series of orders varying with the lens aperture and the angular field. The seven basic aberrations are as follows:

(*a*) *Chromatic.* Since the refractive index of all transparent materials varies with wavelength, all the properties of a lens that depend on refractive index will necessarily vary with wavelength. The simplest of these is the location of an axial image point along the axis. If the position of this image varies with wavelength, the lens is said to suffer from *chromatic aberration.* A lens in which two wavelengths have been united at a common image point is said to be *achromatic,* whereas if three or more wavelengths have been united, it is said to be *apochromatic.*

(*b*) *Lateral color.* A related but quite separate aberration is concerned with a variation with wavelength of the height of an image above the axis, causing colored fringes to surround images far out in the field and to disappear on axis. This aberration is called *lateral color, transverse chromatic aberration,* or *chromatic difference of magnification.*

(*c*) *Spherical.* Spherical aberration exists if the position of an axial image varies from zone to zone along the axis. Even though the marginal and paraxial images may coincide, there may still be a residual of zonal spherical aberration.

(*d*) *Coma.* This is an analogous aberration in which the height of an image above the axis varies from zone to zone of the lens.

It should be noted that these four aberrations are analogous, in the sense that (a) and (b) represent a variation of image position and image height from one wavelength to another, and (c) and (d) represent a variation of image position and image height from one lens zone to another.

(*e*) *Distortion.* Distortion is present in a lens if the height of the image above the axis is not proportional to the height of the object above the axis, i.e., if the magnification varies across the field. If the object is at infinity, distortion is present if the focal length varies with obliquity.

A symmetrical system is such that, if one side is rotated through 180° about the center of the stop, it will fall exactly on the other side. This implies unit magnification also. In such a system, the three transverse aberrations—coma, distortion, and lateral color—are automatically corrected.

(*f*) *Field curvature.* If the image of a plane object perpendicular to the axis does not lie in a plane perpendicular to the axis, then the lens is said to have a *curved field.* This is regarded as inward curving if the image is concave toward the lens, and backward curving if it is convex toward the lens.

(*g*) *Astigmatism.* This is a more subtle aberration in which the image of an off-axis point appears as a pair of focal lines, one line being radial to the field and the other tangential to it. The field curvature will obviously be different for radial and for tangential focal lines. A lens in which astigmatism is corrected at one point in the field is said to be *stigmatic* at that field point, and if that point lies in the focal plane, the lens is said to be an *anastigmat.*

Since, with a distant object, the focal length of a lens is a measure of image size, we can regard coma as a variation of focal length with aperture and distortion as a variation of focal length with obliquity.

V. FIBER OPTICS

An optical fiber is a long thread of glass having an approximately circular section that transmits light along its length by a process of repeated total internal reflection at the surface. No light is lost at total reflection, with the only loss being that due either to absorption in the glass itself or to dirt or grease on the surface of the fiber. To prevent the latter loss it is customary to *clad* the fiber with glass or plastic of a lower refractive index than that of the fiber.

The *numerical aperture* of the fiber is the sine of the slope angle θ of the steepest entering ray that is just at the point of total internal reflection inside the fiber. In Fig. 2.8 it is seen that if ϕ is the critical angle, then $\sin \phi = n_c/n_0$, where n_0 is the refractive index of the fiber and n_c that of the cladding material. The numerical aperture $\sin \theta$ is therefore equal to

$$\sin \theta = n_0 \sin \theta' = n_0 \cos \phi = n_0 \sqrt{1 - \sin^2 \phi} = \sqrt{n_0^2 - n_c^2}.$$

Thus if the index of the fiber is 1.55 and that of the cladding is 1.50, the numerical aperture will be 0.39.

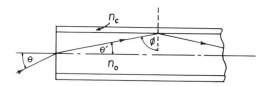

Fig. 2.8. Numerical aperture of a fiber.

A. A Fiberscope

There are many uses to which optical fibers have been put. For example, after laying down an ordered bundle of parallel fibers, the ends can be fused or cemented together and polished to a plane face.[3] An image projected onto one end of the bundle will be transmitted along the bundle in the manner of a long relay system and appear at the other end, where it can be examined through an eyepiece. Such a system has been called a *fiberscope,* and it has been used as a flexible viewing arrangement for examining the inside of mechanical structures and also for medical purposes. The difficulty of fabricating such an ordered bundle is, however, great, and they tend to be very expensive. A broken fiber causes a void to appear in the image. Typically a fiber bundle 5 mm in diameter can be made with fibers of 10 to 20 μm in diameter, and the resolving power of such a bundle is about 35 line pairs/mm.

It is possible to assemble a bundle of ordered fibers and then twist the bundle through 180°, in which case an image projected on one end of the bundle appears inverted when viewed at the other end. Such an invertor can be made extremely short and then cemented together into a rigid unit.

B. Fiber-Optic Face Plates

A large number of ordered fibers can be rigidly fused together and cut into slices, which are fused side-by-side into flat plates about $\frac{1}{8}$ in. thick and several inches in diameter. The plate is then polished on both

---

[3] W. P. Siegmund, Fiber optics, *in* "Applied Optics and Optical Engineering" (R. Kingslake, ed.), Vol. 4, p. 1. Academic Press, New York, 1967.

sides and used as the end window of a cathode-ray tube. The phosphor is deposited on the inside of the plate, and a piece of photographic film can be pressed against the outer face of the window to record the image. The illumination on the film transmitted by such a face plate can be 20 to 40 times as great as when a lens is used to image the phosphor on the film in the ordinary way.

C. Communication Fibers[4]

In the last 10 years or so, means have been found to make fibers in which the absorption loss is around 1 dB/km. [A decibel (dB) is defined as a transmission of $(0.1)^{\frac{1}{10}} = 0.79$; 2 dB is 0.63, and so on, to 10 dB, a transmittance of 0.1.] The real problem with very long fibers is that the time taken for the transmission of a signal via the different possible paths through the fiber varies, so that a rapid sequence of pulses fed into the fiber at one end comes out as a continuous stream at the other end and all information is lost. To prevent this, fibers are being made with a radial gradient of refractive index having a parabolic cross section of index. In this way the light is transmitted by refraction rather than reflection, and in the near infrared the time of transit is the same for all possible paths. The number of possible paths can also be reduced by making the fiber very thin, i.e., of dimensions comparable to the wavelength of the light, in which case the fiber becomes a wave guide. Using light as a carrier wave, a large number of telephone and TV messages can be transmitted over a single fiber with no possibility of interference by ambient electrical disturbances.

D. Selfoc GRIN Rods

Selfoc GRIN rods are a recent development by a Japanese company. A rod 1 or 2 mm thick is made with a radial gradient of refractive index (hence GRIN), and rays of light inside such a rod are curved, oscillating from one side to the other like a sine wave. Such a rod has the property of forming a succession of images wherever the curved rays

---

[4] D. B. Keck and R. E. Love, Fiber optics for communications, *in* "Applied Optics and Optical Engineering" (R. Kingslake and B. J. Thompson, eds.), Vol. 6, p. 439. Academic Press, New York, 1980.

cross each other. These images are successively inverted, then erect, then inverted, etc., at a longitudinal separation of several millimeters. Thus if the rod is cut to the correct length and the ends polished, it acts like a lens, forming an erect or an inverted image as required.

CHAPTER 3

Ray-Tracing Procedures

I. TYPES OF RAYS

Geometrical optics is based on the concept of rays of light, which are assumed to be straight lines in any homogeneous medium and which are bent at a surface separating two media having differing refractive indices. We often need to trace the path of a ray through an optical system, which will generally contain a succession of refracting or reflecting surfaces separated by given distances along the axis. A rough graphical procedure is available for rapid ray tracing, but for more precision it is necessary to use a set of trigonometric formulas executed in succession.

Rays in general fall into three classes: *meridional, paraxial,* and *skew.* Meridional rays lie in the meridian plane, which is the plane containing the lens axis and an object point lying to one side of the axis. If the object point lies on the axis, all rays are necessarily meridional.

Paraxial rays lie throughout their length extremely close to the optical axis. The image formed by paraxial rays is deemed to be *the* image, and if any image departs from the paraxial image, this is regarded as an aberration. Paraxial ray-tracing formulas are purely algebraic, whereas the formulas for tracing real rays involve trigonometric functions.

Skew rays, on the other hand, do not lie in the meridian plane, but they pass in front of or behind it and pierce the meridian plane at the diapoint. A skew ray never intersects the lens axis. Skew rays are much more difficult to trace than meridional rays, and we shall not refer to them again.[1]

---

[1] R. Kingslake, "Lens Design Fundamentals," p. 145. Academic Press, New York, 1978.

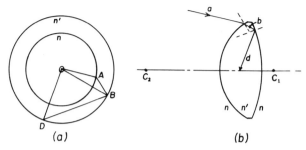

FIG. 3.1. Graphical ray tracing.

II. GRAPHICAL RAY TRACING

There is a simple graphical procedure, based on Snell's construction, by which the path of a meridional ray can be traced through refraction at a lens surface. This procedure, described by Dowell[2] and by van Albada,[3] is as follows: Fig. 3.1b shows a pair of refracting surfaces separating media having refractive indices n and n'. Figure 3.1a shows two concentric circles drawn about a point O having radii proportional to n and n'. To trace a ray, a line OA is drawn through O parallel to the incident ray a up to the circle corresponding to the refractive index n of the medium containing the ray. A line is next drawn from A to B parallel to the normal at the point of incidence, from the n circle to the n' circle. Then OB represents the direction of the refracted ray b. It can readily be seen that this construction is fully in accordance with the law of refraction $n \sin I = n' \sin I'$.

We now proceed to the second surface. In this example, the normal line BD at the second surface is seen to miss the n circle entirely, indicating that total internal reflection will occur, and the reflected ray has the direction d inside the lens.

In practice it is advisable to draw the index circles in ink and to make little tick marks in pencil at the points A, B, D, etc., deleting each mark when the next mark is made, thus displaying only one mark at any one

---

[2] J. H. Dowell, Graphical methods applied to the design of optical systems. *Proc. Opt. Conv.*, **2**, 965 (1926).

[3] L. E. W. van Albada, "Graphical Design of Optical Systems." Pitman, London, 1965.

time. It is unnecessary to draw the radial lines from O to A, to B, etc., for the tick marks are quite sufficient. Changes in the lens can be made at any time by overlaying a sheet of tracing paper and drawing the altered lens upon it. After two or three changes have been made, the original design will have disappeared, and only the changed system will be visible.

The precision of this procedure is not high, say about 1 mm in position and about 1° in direction. The graphical procedure has the great advantage that the lens diameters and thicknesses can be watched continuously as the ray proceeds, to ensure that the dimensions of the system are realistic. Furthermore, the procedure can be operated in reverse, so that we can determine the radius of curvature of a surface that will send a ray into any desired direction. Graphical ray tracing of this kind is perfectly adequate for the design of condensers, magnifiers, and field lenses, but it is not good enough if aberrations are to be determined and corrected.

III. MERIDIONAL RAY TRACING

We define a meridional ray by its slope angle U, which is reckoned positive if a counterclockwise rotation takes us from axis to ray, and by its perpendicular distance Q from the surface vertex. The distance Q is reckoned positive if the ray passes above the surface vertex.

We define a spherical refracting surface by its radius of curvature r, which is considered positive if the center of curvature lies to the right of the surface, and by the refractive indices n and n' of the media lying to left and right of the surface, respectively. The distance measured along the axis from one surface to the next is given by d and is reckoned positive if the light is proceeding from left to right.

The first step in the ray-tracing process is to calculate the angle of incidence I between ray and normal, and this is reckoned positive if a counterclockwise rotation takes us from the normal to the ray. All the data of the incident ray are expressed by plain symbols, and the corresponding data for the refracted ray are given in primed symbols. Figure 3.2a shows that for a spherical surface with radius $r = PC = AC$, the line CN being drawn parallel to the ray

$$Q = r \sin I - r \sin U,$$

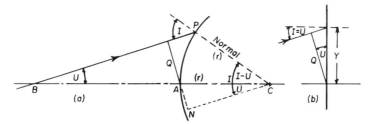

FIG. 3.2. Refraction of a meridional ray: (a) at a sphere; (b) at a plane.

from which

$$\sin I = (Q/r) + \sin U, \qquad (1)$$

or $\sin I = Qc + \sin U$ if the surface curvature c is given instead of its radius r.

We next apply the law of refraction to determine the angle of refraction I':

$$\sin I' = (n/n') \sin I.$$

The third ray-tracing equation is found from the obvious fact that the central angle PCA in Fig. 3.2a is the same for both the entering and emerging rays, or

$$PCA = I - U = I' - U',$$

from which

$$U' = U + I' - I.$$

The final equation is found by adding primes to the first relation, giving

$$Q' = r(\sin I' - \sin U').$$

With these four equations we can determine the U' and Q' of the refracted ray, given the U and Q of the incident ray and the data of the surface: r, n, and n'.

These equations are perfectly general provided that the radius of curvature of the surface is finite. They obviously cannot be applied to a plane surface because then, in the fourth equation, we find that $I' = U'$, and r is infinite, so we have the product of $\infty \times 0$, which is indeterminate. Therefore, for a plane we must develop a separate set of

3. RAY-TRACING PROCEDURES

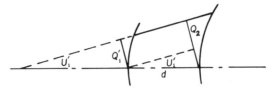

FIG. 3.3. Transfer to the next surface: $Q_2 = Q'_1 + d \sin U'_1$.

equations. From Fig. 3.2b we see that $Y = Q/\cos U = Q'/\cos U'$, and therefore

$$\sin U' = \left(\frac{n}{n'}\right) \sin U \quad \text{and} \quad Q' = Q\left(\frac{\cos U'}{\cos U}\right).$$

In writing a computer program to trace meridional rays, our first act must be to test the value of $c = 1/r$, and if it is zero, we use the plane surface equations, whereas if it is finite, we use the finite radius equations.

In both cases the transfer to the next surface is the same. It can be derived from Fig. 3.3, where we see that

$$Q_2 = Q'_1 + d \sin U'_1.$$

Example. As an example in the use of the ray-tracing equations, we will trace a ray entering parallel to the axis at height 3.172 through the lens shown in Fig. 3.4. This is a typical $f/1.6$ projection lens used for many years for projecting 16-mm and 8-mm movie films in a home projector. In Table I we start by listing the lens data across the page, and follow by the Q and Q' values, and then the angles. The value of the incidence height Y and the sag Z are given in case they are needed. They are found by

$$Y = r \sin(I - U) \quad \text{and} \quad Z = r[1 - \cos(I - U)].$$

IV. PARAXIAL RAYS

A paraxial ray is one which throughout its length lies very close to the lens axis. In theory it should be infinitely close to the axis, but in practice we have some tolerance, and we consider a ray as being

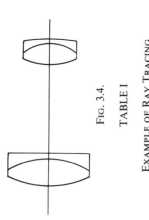

Fig. 3.4.

TABLE I

EXAMPLE OF RAY TRACING

r		8.572		−7.258		5.735		−3.807		−16.878
c		0.1166589		−0.1377790		0.1743679		−0.2626740		−0.0592487
d			2.4		7.738		1.8		0.4	
n	1.0		1.52240		1.61644		1.51625		1.61644	1.0
				Marginal ray $f/1.6$						
Q		3.172	2.905252		2.902741	1.665901		1.299741		1.254617
Q'		3.224772	2.941579		2.880377	1.663256		1.319817		1.200372
I		21.71821	−32.23657		∞	7.67363		−32.91271		−13.72930
I'		14.06750	−30.15782			5.05236		−30.64266		−22.55920
U	0.0		−7.65070		−5.57196		−9.02988		−11.65115	−9.38110	−18.21100
Y		3.172	3.020		2.917	1.648		1.381		1.280
Z		0.608	−0.658		0	0.242		−0.259		−0.049
				Paraxial ray						
y		1.0	0.903927		0.891744	0.510797		0.397768		0.376798
u	0.0		−0.040031	−0.030456		−0.049231	−0.062794		−0.052426	−0.098505

Marginal $L' = 3.840978$, paraxial $l' = 3.825163$, and focal length = 10.151767.

3. RAY-TRACING PROCEDURES

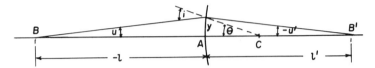

FIG. 3.5. Paraxial ray at a surface.

"paraxial" as long as the aberration is negligible within the degree of precision of our work.

We cannot use the ordinary ray-tracing equations for a paraxial ray because the sines are not given for extremely small angles. However, we know that for very small angles the sine is equal to the angle itself in radian measure, so we can reduce the ordinary ray-tracing formulas to their limiting case where $\sin U$ can be replaced by u and $\sin I$ by i. Lowercase letters are used for paraxial data to remind us that they are infinitesimals.

A paraxial ray is shown in Fig. 3.5, greatly exaggerated in all vertical dimensions. Each quantity is indicated with its correct sign. The ray enters the surface from the left at an infinitesimal slope u and strikes the surface at an infinitesimal height y. The normal at the point of incidence is shown dashed, and it passes through the center of curvature of the surface at C. The angle of incidence between the ray and the normal is i, also infinitesimal, and the central angle is θ. Evidently θ is equal to $i - u$, and because the ray is infinitely close to the axis, $y = r\theta$, where the angle θ is expressed in radians. Hence

$$i - u = y/r \quad \text{or} \quad i = (y/r) + u. \quad (2)$$

This equation is the paraxial equivalent of the finite Eq. (1) given in Section III.

The data of the refracted ray are similar to those of the incident ray, with all ray data being primed. Hence

$$i' = (y/r) + u'. \quad (3)$$

Multiplying Eq. (2) by n and Eq. (3) by n' gives two expressions which are equal, because the law of refraction for paraxial rays is merely $ni = n'i'$. Therefore

$$n'\left(\frac{y}{r} + u'\right) = n\left(\frac{y}{r} + u\right) \quad \text{or} \quad (nu)' = (nu) - y\left(\frac{n' - n}{r}\right). \quad (4)$$

This is a relation between the slope of the entering paraxial ray and the slope of the emerging ray for a single refracting surface. Note that the angles of incidence and refraction have disappeared; they are merely auxiliaries and need not be calculated.

Another important relation is also shown in Fig. 3.5. If l and l' are the distances from the surface to the points where the incident and refracted rays cross the lens axis, then

$$y = -lu = -l'u'; \quad \text{hence} \quad u = -(y/l) \quad \text{and} \quad u' = -(y/l').$$

The quantities l and l' are called the *intersection lengths* of the ray, and they have the usual signs of negative to the left of the surface and positive to the right.

If we divide Eq. (4) by y we obtain

$$n'\left(\frac{u'}{y}\right) = n\left(\frac{u}{y}\right) - \left(\frac{n' - n}{y}\right),$$

from which

$$\frac{n'}{l'} = \frac{n}{l} + \frac{n' - n}{r}. \tag{5}$$

Now all the paraxial angles have disappeared, and they are all merely auxiliaries. This relation shows that all paraxial rays emerging from a given object point pass through the same image point. This is certainly not true of finite rays. We can readily calculate the value of l', given l and the surface data, by

$$l' = \frac{n'}{(n/l) + \phi},$$

where ϕ is the *surface power*, given by $(n' - n)/r$. A small pocket calculator equipped with a reciprocal key is a convenient means for using this equation. The transfer to the next surface is extremely simple, because $l_2 = l'_1 - d$.

A. Paraxial Ray with All Angles

There are, of course, other ways to trace a paraxial ray. For instance, we can trace a paraxial ray with all the angles by using these equations in order: given the l and y of the incident ray, we have

$$u = -y/l,$$
$$i = (y/r) + u,$$
$$i' = (n/n')i,$$
$$u' = i' - (y/r),$$
$$l' = -y/u',$$

with the transfer

$$l_2 = l'_1 - d.$$

These equations can be collected together to give

$$u' = \left(\frac{y}{r} + u\right)\frac{n}{n'} - \frac{y}{r},$$

or

$$u' = (cy + u)\frac{n}{n'} - cy, \quad \text{where} \quad c = \frac{1}{r}.$$

The transfer is now

$$y_2 = y_1 + du'_1. \tag{6}$$

B. The y-nu Method for Tracing Paraxial Rays

In many ways the most useful procedure for tracing paraxial rays is to use Eq. (5) for the refraction and Eq. (6) for the transfer. Both these expressions are of the same form, namely, "the new value is equal to the old value plus the product of the other variable times a constant." The truth of this can be seen by

$$(nu)' = (nu) + y(-\phi),$$
$$y_2 = y_1 + (nu)'_1(d/n).$$

To trace a ray by this method, we start by tabulating the constants $-\phi$ and d/n across the page, followed by the values of y and nu in order, calculated and recorded in a zig-zag fashion, with the initial y and nu being given. The calculation ends with the determination of the image

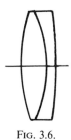

FIG. 3.6.

distance l' by

$$\text{image distance} = -\frac{\text{last } y}{\text{last } u'}.$$

As an example, we will use this procedure to trace a paraxial ray from infinity through the cemented doublet lens shown in Fig. 3.6. The figures to be recorded are shown in Table II. As before, any data referring to the surfaces are written in the columns and data referring to the spaces between surfaces are written between the columns. Having entered the values of y_1 and $(nu)_1$, we find $(nu)'_1 \equiv (nu)_2$ by

TABLE II

TRACING OF A PARAXIAL RAY FROM INFINITY THROUGH A CEMENTED DOUBLET LENS

			r		7.3895		−5.1784		−16.2225		
			d			1.05		0.4			
			n	1.0		1.517		1.649		1.0	
$-\phi = (n - n')/r$					−0.0699641		0.0254904		−0.0400061		
			d/n			0.6921555		0.2425712			
			y		2.0		1.903148		1.880973		$l' = 11.28584$
			nu	0		−0.1399282		−0.0914162		−0.1666667	$f' = 12.0$
			u	0		−0.0922400		−0.0554373		−0.1666667	
			y/r		0.2706543		−0.3675166		−0.1159484		
$i = (y/r) + u$					0.2706543		−0.4597566		−0.1713857		
			l		∞		20.63255		33.92977		
			l'		21.68255		34.32977		11.28585		

adding the product of y_1 and the number over it, namely, $-\phi_1$, to the given value of $(nu)_1$. This simple process is repeated twice for each surface in the lens, once for the refraction and once for the transfer. Any starting data can be used provided that the y is equal to the product of l and $-u$; some workers always use $y_1 = 1.0$.

V. CURVED MIRRORS

A concave or convex reflecting surface can be included in a lens system when tracing rays. The rule is to list the surfaces in succession in the order in which the light strikes them, with the correct axial separation d and refractive index n between the surfaces (Table III). Note particularly that if the light is traveling from right to left, both the separation and the refractive index must be entered with a negative sign. Other than this there is no distinction between a lens and a mirror. An example of a paraxial ray traced through the catadioptric system is shown in Fig. 3.7.

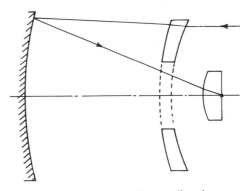

FIG. 3.7. Ray tracing through a catadioptric system.

VI. MAGNIFICATION AND THE LAGRANGE THEOREM

An *image magnification* is the transverse dimension of the image divided by the transverse dimension of the object. For a single spheri-

TABLE III

Tracing a Paraxial Ray through a Catadioptic System[a]

r	5.0		6.993		10.0		1.645		∞		
d		-0.35		-4.0		5.286		0.6			
n	-1.0		-1.545		-1.0		1.0		1.545		1.0
$-\phi$	0.109000		-0.0779350		-0.2		-0.3313066		0		$l' = 0.00006$
d/n		.2265372		4.0		5.286		0.3883495			$f' = 4.37605$
y	2.0		2.049385		2.282510		0.177515		0.000026		
nu	0.0	0.218000		0.0582812		-0.3982208		-0.4570327			-0.4570327

[a] The sign of nu depends both on the sign of n and the sign of u. Otherwise the sign of u remains the same as usual, namely, that a ray sloping downwards from left to right has a negative value of u.

cal surface it is easy to develop a formula for the paraxial magnification, and this can readily be extended to apply to a complete system.

In Fig. 3.8, suppose that an axial object point A is imaged at A' by a single refracting surface. Then a small object of height h erected at A will be imaged at A' with height h'. The magnification is, of course, $m = h'/h$.

To relate h and h', we trace a paraxial ray from B, the top of the object, to the surface vertex and on to the top of the image at B'. The angles of incidence and refraction of this ray are θ and θ', respectively. Therefore

$$\theta = -h/l \quad \text{and} \quad \theta' = -h'/l'.$$

We now multiply θ by yn and θ' by yn'. Then, because the law of refraction for paraxial rays is $n\theta = n'\theta'$, we obtain

$$nh(y/l) = n'h'(y/l'),$$

which can be further simplified since $y/l = -u$ and $y/l' = -u'$, giving finally

$$h'n'u' = hnu.$$

This relation is known as the *theorem of Lagrange*.

For a complete system, it is clear that n'_1 is identical with n_2, h'_1 is identical with h_2, u'_1 is identical with u_2, etc., throughout the system. We therefore conclude that the product hnu is invariant for all the spaces between surfaces, including the object space and the image

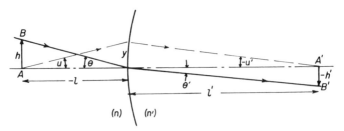

FIG. 3.8. The Lagrange relationship.

space. This product is often called the *Lagrange invariant,* or more often the *optical invariant.* Its value depends on the height of the object and the particular choice of paraxial ray.

Clearly, if we have traced a paraxial ray through a lens by the $y-nu$ method, the image magnification is merely equal to the ratio of the starting value of nu to the emerging value of $(nu)'$. If the lens is in air, as most lenses are, the image magnification becomes merely

$$m = u_1/u'_k,$$

assuming that there are k surfaces in the lens. No such simple relationship is available if the ray is traced by the (l, l') method. Consequently, if the magnification is required to be known, it is essential that we trace the ray by the $y-nu$ method.

An important result of the Lagrange theorem is that if by some means we can vary the aperture of a system, then the angular field must change in inverse ratio to the aperture. Briefly, we say that the product of aperture and field is constant and that increasing one must result in a reduction of the other.

VII. THE FRESNEL LENS

A *Fresnel lens* is a thin, flat, usually plastic lens, which is plane on one side and has sloping rings on the other, each ring being at such a slope as to refract a ray from a given object point to a specified image point.

If A and B are the distances of object and image from the lens and Y is the radius of a ring on the lens, then the slopes of the ray upon entering and leaving the lens are given by

$$\tan a = Y/A \quad \text{and} \quad \tan b = Y/B.$$

VII. THE FRESNEL LENS

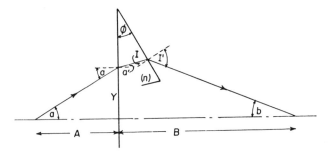

FIG. 3.9. Illustrates one ring of a Fresnel lens.

If the slope of the ring on the lens is ϕ, as shown in Fig. 3.9, and the refractive index of methacrylate plastic is $n = 1.4917$, we see that the ray enters the plane side of the lens at an angle of incidence a, giving the ray slope inside the lens as a', where $\sin a' = (\sin a)/n$. Then the angles of incidence and refraction at the sloping surface are respectively,

$$I = \phi - a' \quad \text{and} \quad I' = \phi + b.$$

Since $\sin I' = n \sin I$, we see that

$$\sin \phi \cos b + \cos \phi \sin b = n \sin \phi \cos a' - n \cos \phi \sin a'.$$

Dividing through by $\cos \phi$ gives the solution for the lens slope as

$$\tan \phi = \frac{\sin a + \sin b}{n \cos a' - \cos b}$$

because $n \sin a' = \sin a$.

Example. Suppose $A = 24$ in. and $B = 20$ in. Then for a ring of radius $Y = 6$ in. we have

$$a = 14.0362°; \quad a' = 9.3573°; \quad \text{and} \quad b = 16.6992°.$$

Plugging these into the expression for ϕ gives $\tan \phi = 1.0385$ from which $\phi = 45.8704°$.

A metal mold is made up on the basis of these calculations, and plastic lenses are pressed from the mold. Care must be taken to ensure that the rings are close enough to one another so as not to be visible in the intended application of the Fresnel lens. Also, the wider the rings, the thicker the finished lens must be.

CHAPTER 4

The Gaussian Theory of Lenses

I. INTRODUCTION

Prior to the time of C. F. Gauss (1777–1855), the term *focal length* had meaning only for a very thin lens, it being the distance from the lens to the "focus" where images of distant objects were formed. Before the invention of photography in the 1830s, all practical lenses were thin enough for this concept of focal length to be quite adequate. However, with the introduction of the Petzval Portrait lens in 1839, some extension of this concept became necessary.

II. THE FOUR CARDINAL POINTS

Gauss met the problem of focal length for thick lenses by postulating four cardinal points in any lens, two focal points and two principal points, which he defined in the following manner. Figure 4.1 shows a lens of any construction, with a set of rays A, B, C, etc., entering the left-hand end of the lens parallel to the lens axis. After passing through the lens these rays cross the axis at various points H, J, etc., in the image space, and by extending each ray backward or forward as needed until the entering and emerging portions intersect, we can locate an *equivalent refracting point* for each ray, with the locus of all such points Q, R, etc., being the *equivalent refracting locus* of the lens for parallel rays entering from the left. The portion of this locus within the paraxial region is a plane perpendicular to the lens axis called the *principal plane* P_2 and the image point for paraxial rays lies at the *focal point* F_2. The axial distance from the principal point to the focal point is called the *focal length* of the lens. (At one time it was called the equivalent focal length, meaning the focal length of the equivalent thin lens, but the word "equivalent" is quite unnecessary.)

There is another pair of principal and focal points for light entering

II. THE FOUR CARDINAL POINTS

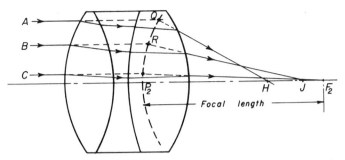

FIG. 4.1. The equivalent refracting locus of a lens for a set of rays entering from the left parallel to the axis.

the lens parallel to the axis from the right and emerging to the left. The complete set of four cardinal points F_1, P_1, P_2, and F_2 is shown in Fig. 4.2.

A. Relation between Principal Planes

Working purely from the formal definitions for the cardinal points, we can derive several relations which are of considerable importance in geometrical optics. The first of these is that the two principal planes are images of each other at unit magnification (see Fig. 4.2). In this diagram a paraxial ray A enters from the left parallel to the axis. It is effectively bent at Q in the second principal plane (following Gauss's definitions), and it emerges through the second focal point F_2. Another paraxial ray B enters parallel to the axis from the right in the same straight line as the first ray. This second ray is effectively bent at R in the P_1 plane and emerges through the first focal point F_1. By reversing the arrows on this second ray, we end up with two rays entering through R and emerging through Q. This makes Q an image

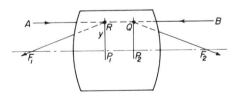

FIG. 4.2. The principal planes as unit planes.

of R and vice versa. In Fig. 4.2 both the object and image points are virtual, but other systems could be drawn in which one or both of the principal points fell outside the system instead of inside. Because R and Q are at the same height above the axis, the magnification for this conjugate pair is $+1.0$. For this reason the principal planes are often called the *unit planes* of a lens. It must be emphasized that the cardinal points and focal length are all paraxial properties of a lens.

B. Relation between the Focal Lengths

We can now proceed to determine the relation between the two focal lengths of a lens system. Suppose we place a small object at the anterior focal point F_1, as shown in Fig. 4.3, and draw two rays from the top of this object into the lens. The first ray enters parallel to the axis and emerges through F_2 at the right. The second ray enters the lens toward the first principal point P_1, and therefore it must emerge through P_2, because P_2 is an image of P_1. How about the direction of this emerging ray? Since the small object h is located in the anterior focal plane, all rays originating at any point in the object must emerge as a parallel beam at the other end of the lens, because the focal plane is conjugate to infinity. Consequently, our second ray must emerge through P_2 parallel to the emergent portion of the first ray. If the slope of the second ray on entry is ω and on emergence is ω', as shown in Fig. 4.3, we see that

$$\omega = -h/f \quad \text{and} \quad \omega' = h/f',$$

from which $\omega'/\omega = -f/f'$.

To obtain a second expression for ω'/ω, we move the small object along the axis until it falls on the anterior principal plane at P_1. The image is now at P_2, and the magnification is unity. Applying the Lagrange equation to this pair of conjugates we see that

$$hn\omega = hn'\omega',$$

hence $\omega'/\omega = n/n'$, where n and n' are the refractive indices of object space and image space, respectively. Combining the two values of ω'/ω gives

$$-f/f' = n/n' \quad \text{or} \quad f = -f'(n/n').$$

For a lens in air, therefore, the two focal lengths are equal in magnitude

II. THE FOUR CARDINAL POINTS

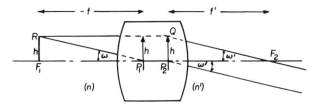

FIG. 4.3. Ratio of the two focal lengths.

and opposite in direction, so that if one focal point lies to the left of its principal point, the other focal point must lie to the right of its principal point.

When we speak of *the* focal length of a lens we always refer to the f' value, which is positive for a positive (convex) lens and negative for a negative (concave) lens. The other focal length f has the opposite sign, which is confusing, and therefore it is best to use only the f' value in all formulas relating to a lens.

It should be clear from the foregoing discussion that if we trace a ray from left to right entering parallel to the axis, the focal length and back focal distance will be given by

$$f' = -\frac{\text{entering ray height}}{\text{emerging ray slope}} = -\frac{y_1}{u'_k},$$

$$l' = -\frac{\text{emerging ray height}}{\text{emerging ray slope}} = -\frac{y_k}{u'_k}.$$

The distance of the second principal point from the rear vertex of the lens is found by

$$l'_{pp} = l' - f'.$$

A ray entering a lens from the front strikes the first principal plane and then jumps across to the second principal plane at the same height before continuing on its way. For this reason the space between the two principal planes is often called the *hiatus*.

C. THE LAGRANGE EQUATION FOR A DISTANT OBJECT

In its simple form, the Lagrange equation $hnu = h'n'u'$ has no meaning for an infinitely distant object, for then $u = 0$ and $h = \infty$, and the product of zero times infinity is indeterminate.

4. THE GAUSSIAN THEORY OF LENSES

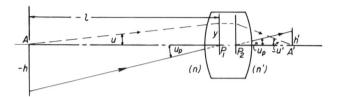

FIG. 4.4. The Lagrange relationship for a distant object.

To interpret the Lagrange equation for a distant object, we refer to Fig. 4.4. The point A represents an object point that is distant but not infinitely distant. Then

$$h'n'u' = (h/l)n(lu).$$

In the limit when l becomes infinite, the ratio h/l becomes equal to $\tan u_p$ and the product lu becomes equal to $-y$, where y is the height at which the ray enters the lens. Thus, since $y/u' = -f'$, we find

$$h' = -(n/n') \tan u_p (y/u') = (n/n')f' \tan u_p,$$

where u_p is the slope of an entering parallel beam of light. For a lens in air, therefore, the image height for an infinitely distant object is equal to the product of the focal length and the tangent of the angular subtense of the object. Paraxially, the tangent is equal to the angle in radians.

Another way of looking at this is seen in Fig. 4.5, where the anterior focal length $f = -(n/n')f'$, hence

$$h' = -f \tan u_p.$$

A ray entering at slope u_p through the anterior focal point emerges parallel to the axis at a height h' on the other side of the lens.

We thus have two expressions for focal length, each of which consists of a length divided by an angle. They are

$$f' = -y_1/u' \quad \text{and} \quad f' = h'/u_p.$$

The first of these relationships is used to compute the focal length after tracing a paraxial ray, whereas the second expression provides a simple and unambiguous definition of focal length. This will be used later in Section VII.

Both these formulas refer to paraxial rays. If the focal length varies

II. THE FOUR CARDINAL POINTS

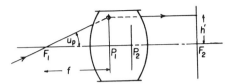

FIG. 4.5. Illustrating that $h' = f \tan u_p$.

with the lens aperture, the resulting aberration is coma. If the focal length varies with obliquity, then the resulting aberration is distortion.

D. LENS POWER

The *power* of a lens or surface is basically the reciprocal of its focal length. To be specific, the power of a single refracting surface is defined as $\phi = (n' - n)/r$, where n is the refractive index on one side of the surface, n' that of the other side, and r the radius of curvature of the surface. The power of a complete lens system is defined as n'/f', where f' is the focal length and n' the refractive index of the image space. It is, of course, also equal to $-n/f$ in terms of the anterior focal length and the refractive index of the object space, although the latter meaning is seldom used. For a thin lens the usual unit of power is the *diopter*, which is the reciprocal of the meter; hence a thin lens having a focal length of 20 cm has a power of 5 diopters. Of course, any desired units can be used for focal length and for power, but the diopter is well established, particularly in the field of spectacle lenses.

We can determine the contribution of any one surface within a lens to the power of the lens by the following argument. By Eq. 4 of Chapter 3 we see that

$$\text{for surface (1):} \quad (nu)'_1 - (nu)_1 = -y_1\phi_1,$$
$$\text{for surface (2):} \quad (nu)'_2 - (nu)_2 = -y_2\phi_2,$$
$$\vdots$$
$$\text{for surface }(k): \quad (nu)'_k - (nu)_k = -y_k\phi_k.$$

When we add all these expressions together, extensive cancellation occurs because $(nu)'_1 \equiv (nu)_2$, etc., so that the sum becomes merely

$$(nu)'_k - (nu)_1 = \sum -y\phi.$$

Dividing through by y_1 gives

$$\frac{(nu)'_k}{y_1} - \frac{(nu)_1}{y_1} = \Sigma - \frac{y}{y_1}\phi.$$

So far the expression is perfectly general. We now consider the special case of an infinitely distant object, for which u_1 is zero. Furthermore, for such an object $(-nu)'_k/y_1$ is the lens power P. Hence the lens power is given by

$$P = \Sigma\left(\frac{y}{y_1}\cdot\phi\right) = \Sigma\frac{y}{y_1}\left(\frac{n'-n}{r}\right).$$

This tells us that the contribution of a single surface to the power of the lens is equal to the surface power ϕ multiplied by the ratio of the height of the ray at the surface divided by its height of entry into the lens. If a ray drops before striking a surface, that surface contributes less than its own power to the power of the system, and conversely, if the ray rises at the surface, that surface contributes more than its power to the system. In the limit when $y = 0$ at a surface, that surface makes no contribution to the lens power, whether it is convex, concave, or plane. Indeed, it could well be made of ground glass, because the image falls on the surface itself in such a case.

This law leads to a ready explanation of many effects in lenses. For instance, it shows why increasing the thickness of a biconvex lens reduces its power and increases its focal length, but increasing the thickness of a meniscus lens reduces the negative contribution of the concave surface and therefore *increases* the lens power.

It should be noted particularly that if the image is formed in a material other than air, the focal length is still given by $f' = -y_1/u'_k$, but the lens power is now equal to n'/f'. Only when the image is formed in air is the power equal to the reciprocal of the focal length. However, this is by far the most common situation, and in very few instances is the image formed in a material other than air.

Do not be confused by the appearance of a lens in a drawing. Just because it is thick does not make it strong. Figure 4.6 shows three lenses which all have the same focal length. The first is thin and small, the second is large and thick, and the third is an ordinary achromatic doublet. As a good example of this effect, turn ahead to Figure 12.5, in which five eyepieces of different types are all drawn to the same focal length, yet example (e) looks as though it must be much stronger than example (a) because it is thicker and larger.

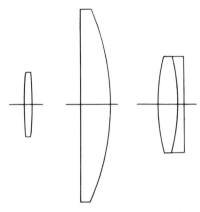

FIG. 4.6. Three lenses all drawn to the same focal length.

III. CONJUGATE DISTANCE RELATIONSHIPS

Having determined the locations of the four cardinal points of a lens, we can find at once the position and size of the image of a given object. The distances of object and image can be measured either from the focal points or from the principal points, depending on which is most convenient.

A. Distances from Focal Points

If the object AB in Fig. 4.7 is located at a distance z from the F_1 point (considered negative if the object is to the left of the focal point) and if the image is at a distance z' from the F_2 point, then two rays can be drawn from the point A at the top of the object; one entering the lens parallel to the axis (which will be effectively bent at Q_2 and emerge through the focal point F_2) and the other ray entering the lens through F_1, (being effectively bent at Q_1 and emerging parallel to the axis). The intersection of the emerging portions of these two rays locates the image of A at A'. By the similar triangles ABF_1 and $F_1P_1Q_1$, we see that

$$h/-z = -h'/-f \quad \text{or} \quad m = h'/h = -f/z, \tag{1a}$$

4. THE GAUSSIAN THEORY OF LENSES

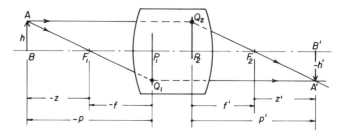

FIG. 4.7. Conjugate distances from a lens.

and by similar triangles $Q_2P_2F_2$ and $F_2B'A'$, we see that

$$h/f' = -h'/z' \quad \text{or} \quad m = h'/h = -z'/f'. \tag{1b}$$

Since the magnification $m = -f/z = -z'/f'$, we can cross multiply, giving

$$zz' = ff'. \tag{2}$$

Since, for any lens, $f = -f'(n/n')$, we see that this equation can be written as

$$zz' = -f'^2(n/n').$$

For a lens in air, $n = n' = 1$, so this equation simply becomes

$$zz' = -f'^2.$$

This relation has many uses. For example, we can use it to lay out the focusing scale on a camera lens. If the object is at a distance p from the front principal plane of the lens, then $z = p - f'$, and the distance through which the lens must be moved from the infinity setting to focus on this object is given by

$$z' = f'^2/(p - f').$$

Another application forms the basis of the optician's lensometer. This is an instrument for measuring the power of a spectacle lens. In Fig. 4.8 the lens A under test is placed at the anterior focus of a known lens B built into the instrument, and a parallel beam of light is admitted to it. With no added lens the image is formed at F_2. When an eyeglass lens is inserted at A the object for lens B is distant from its anterior focus by $z = f'_A$, and hence the image distance from F_2 is given by $z' = -f'^2_B/f'_A$. The image is thus moved longitudinally by a distance

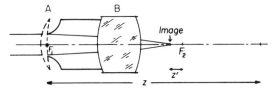

FIG. 4.8. The structure of an optician's lensometer.

proportional to the power of the added lens, to the left if the added lens is positive and to the right if it is negative. In practice, the light is made to traverse this system backward, with a movable illuminated reticle at F_2 and a small telescope enabling the operator to tell when the emerging light is parallel.

Strictly speaking, the object distance for lens B is equal to the back focal distance of the eyeglass lens and not its focal length. However, this is what the optician needs to know, and for this reason the instrument is sometimes called a *vertometer* because it determines the vertex focal distance of the lens.

B. Distances from Principal Points

If the locations of the principal points are known or can be assumed, then it is often more convenient to relate the conjugate distances of object and image from the principal points rather than the focal points. If these distances are represented by p and p', respectively, we see that

$$z = p - f \quad \text{and} \quad z' = p' - f'.$$

Equation (2) now becomes

$$(p - f)(p' - f') = ff'$$

from which

$$(f'/p') + (f/p) = 1.$$

Because $f = -f'(n/n')$, this formula can be simplified to

$$\text{lens power} = n'/f' = n'/p' - n/p.$$

For a lens in air, which is of course the usual situation, we have the familiar relationship

$$1/p' - 1/p = 1/f'. \tag{3a}$$

The image magnification is found from Eq. (1) to be

$$m = p'n/pn' \qquad (3b)$$

or, for a lens in air, $m = p'/p$.

C. Nodal Points

The nodal points of a lens are a pair of conjugate points having the property that a paraxial ray entering toward the first will emerge from the second at the same slope. In Fig. 4.9 we see a ray entering toward the first nodal point N_1 at a slope ω. Applying the Lagrange theorem to an object located at N_1 we find that

$$h'n'\omega = hn\omega,$$

from which the magnification for this pair of conjugates is given by

$$m = h'/h = n/n'.$$

Substituting this into Eq. (1) tells us that the distance from F_1 to N_1 is equal to the posterior focal length f', and the distance from N_2 to F_2 is equal to the anterior focal length f. The whole array of six cardinal points is shown in Fig. 4.9. Of course, if the lens is in air, N_1 coincides with P_1 and N_2 with P_2, and there is nothing to distinguish between the nodal and the principal points. However, the nodal points are important when studying the optical system of the eye, because the refractive index of the image space is then about 1.34 and the two focal lengths are unequal. In fact, the nodal points were first proposed by an ophthalmologist J. B. Listing because he needed them in his work.

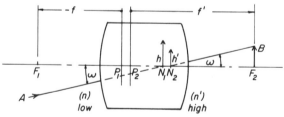

Fig. 4.9. The nodal points of a lens.

IV. A SINGLE LENS

A. A Single Thick Lens

We can easily derive useful general formulas for the focal length and back focal distance of a single thick lens by writing Eqs. (4) and (6) of Chapter 3 in algebraic terms (see Table I).

TABLE I

RAYTRACE THROUGH A SINGLE THICK LENS

		r_1		r_2	
r d n	1.0		t N		1.0
$-\phi$		$\dfrac{1-N}{r_1}$		$\dfrac{N-1}{r_2}$	
$\dfrac{d}{N}$			$\dfrac{t}{N}$		
y		1.0		$1 + \dfrac{t}{N}\left(\dfrac{1-N}{r_1}\right)$	
u	0.0		$\dfrac{1-N}{r_1}$		$\dfrac{1-N}{r_1} + \dfrac{N-1}{r_2}\left[1 + \dfrac{t}{N}\left(\dfrac{1-N}{r_1}\right)\right]$

Hence the focal length is given by

$$\frac{1}{f'} = (N-1)\left[\frac{1}{r_1} - \frac{1}{r_2} + \frac{t}{r_1 r_2}\left(\frac{N-1}{N}\right)\right], \qquad (4)$$

and the back focus by

$$\mathrm{BF} = f'\left[1 - \frac{t}{r_1}\left(\frac{N-1}{N}\right)\right].$$

The rear principal plane is therefore located at a distance from the rear lens vertex given by

$$l'_{\mathrm{pp}} = \mathrm{BF} - f' = -\frac{tf'}{r_1}\left(\frac{N-1}{N}\right).$$

To locate the anterior principal plane we reverse the direction of the ray, giving

$$\text{FF} = f\left[1 + \frac{t}{r_2}\left(\frac{N-1}{N}\right)\right],$$

hence

$$l_{pp} = \text{FF} - f = \text{FF} + f' = -\frac{tf'}{r_2}\left(\frac{N-1}{N}\right).$$

The separation between the two principal planes is often called the *hiatus*, represented by the symbol Z. Its value is found by

$$Z = t - l'_{pp} + l_{pp} = t(N-1)\left[\frac{(r_1 - r_2) - t}{N(r_1 - r_2) - t(N-1)}\right].$$

If a lens is fairly thin, we may neglect the t inside the bracket, and in this case

$$Z \simeq t\left[\frac{N-1}{N}\right].$$

Thus, if $N = 1.5$, the hiatus is approximately equal to one-third the thickness of the lens.

B. A Thin Lens

A strictly "thin" lens, i.e., one having zero thickness, is of course a purely mathematical fiction. However, the simplicity and convenience of the thin-lens concept is so great that we often speak of "thin lenses" as being those in which the thickness effects are negligible within the precision of our work. Now we assume that the principal planes coalesce within the lens itself, and the conjugate distances are measured "from the lens." A thin lens is specified completely by its focal length and diameter, whereas if its shape is important, we add the radii of curvature and the refractive index.

Writing $t = 0$ in Eq. (4) tells us that the power of a thin lens is given by

$$\frac{1}{f'} = (N-1)\left(\frac{1}{r_1} - \frac{1}{r_2}\right) = \phi_1 + \phi_2,$$

where the ϕ's are the surface powers. Because the ray height y is the same at both surfaces in a thin lens, we merely add the surface powers to determine the lens power. We cannot do this if the lens has a finite thickness.

C. Simplified Case of a Thin Lens with Real Conjugates

The most common situation in real life by far is that in which a positive lens is used to form a real image of a real object. For this special case we can simplify Eq. (3) by ignoring signs and writing the object and image distances as a and b and the magnification as m, with all three quantities being regarded as simple positive numbers. When we do this we find that

$$f' = \frac{ab}{a+b} = \frac{\text{product}}{\text{sum}}; \qquad m = \frac{b}{a};$$

$$a = f'\left(1 + \frac{1}{m}\right); \qquad b = f'(1 + m);$$

and

$$D = a + b = f'\left(2 + m + \frac{1}{m}\right).$$

The two middle equations are read: "Object distance is 1 plus $1/m$ focal lengths" and "Image distance is 1 plus m focal lengths." Familiarity with these relations enables one to make rapid mental calculations of image distances and magnification.

For example, if a 4-in. lens is required to form an image at a magnification of six times, these formulas tell us that the object distance must be $a = 4 \times 7/6 = 4.67$ in. and the image distance $b = 4 \times 7 = 28$ in. Similarly, if we know that the object and image distances are, respectively, 30 and 5 in., we see that the focal length of the lens must be $(30 \times 5)/(30 + 5) = 4.29$ in. We can use these equations for a thick lens if we remember that the distances a and b must be measured from the two principal planes.

The last of these formulas is particularly useful in designing a printer to produce prints of a standard size from a variety of negative sizes. Thus, knowing that the object-to-image distance is D and the required magnification is m, we can at once determine the required focal length by

$$f' = \frac{D}{2 + m + (1/m)}. \tag{5}$$

If we are using a thick lens, we must remember to subtract the hiatus from the value of D before making the calculation.

It is sometimes useful, for example in the design of some types of zoom lenses, to invert Eq. (5) and solve for the magnification that will result when a lens of known focal length is used between given object and image planes. Then

$$m = \tfrac{1}{2}k - 1 \pm \sqrt{\tfrac{1}{4}k^2 - k}, \quad \text{where} \quad k = D/f'.$$

Then, when f' and m are known, the distances a and b can be found by the ordinary formulas. It should be noticed that the two values of m given by this formula are reciprocals, so that when a given lens is used between fixed object and image planes, the lens may occupy either of two positions, giving magnifications equal to m and $1/m$, respectively. This feature has been used in some types of variable-power telescopes (see Chapter 12, Section III.G).

D. Insertion of Thickness into a Thin Lens

Suppose we have designed a system that includes a thin lens of power ϕ, refractive index N, and two surface curvatures c_1 and c_2. We now wish to replace this by a thick lens of thickness t, having the same focal length and the same ratio of radii $k = r_2/r_1 = c_1/c_2$. For the thin lens, the surface curvatures were

$$c_2 = \frac{\phi}{(N-1)(k-1)}; \quad c_1 = kc_2.$$

For the thick lens we have

$$\phi = (N-1)\left[c_1' - c_2' + \frac{t}{N}(N-1)c_1'c_2'\right] \quad \text{with} \quad k = \frac{c_1'}{c_2'}.$$

The solution to this quadratic is

$$c_2' = \frac{(1-k) \pm \sqrt{(k-1)^2 + 4kt\phi/N}}{2kt(N-1)/N}; \quad c_1' = kc_2'.$$

IV. A SINGLE LENS

As an example, suppose $\phi = 0.1$, $N = 1.523$, and $k = -1.5$. This represents a biconvex lens with the front surface stronger than the rear surface. For the thin lens, the solution is $c_1 = 0.1147$ and $c_2 = -0.0765$. If we now insert a thickness of 0.2, the solution yields the curvatures $c_1' = 0.1151$ and $c_2' = -0.0767$. It will be seen that both surfaces have become slightly stronger as a result of the insertion of thickness, but only slightly. Thickness has very little effect on the power of a biconvex lens, but it has much greater effect if the lens is meniscus.

E. A Collimator

It has been pointed out that the focal plane of a lens is conjugate to infinity, so that the image of a very distant object falls within the focal plane. Conversely, an object located within the focal plane is imaged at infinity. Such a system is called a *collimator.*

If the source of light in a collimator is a single point, a parallel beam will emerge from the other side of the lens, but it will contain very little energy. However, if the object has finite dimensions, a bundle of parallel beams will emerge from the other side of the lens, diverging from one another (Fig. 4.10). Thus the so-called parallel beam of light from a collimator is actually an expanding beam, with the angle of divergence being given by the ratio of the source size to the focal length of the lens. For this reason, it is much better to call this a collimated beam rather than a parallel beam.

If the source happens to have the same diameter as the lens, the entering beam on the object side will have a parallel external form, but each elementary beam entering the lens will be divergent. Such a beam

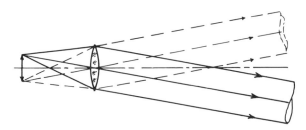

Fig. 4.10. A collimator forms an expanding bundle of parallel rays.

is called a *quasi-parallel* beam. We thus have the paradox that a parallel beam has a divergent external form, whereas in this case a diverging beam has a parallel external form. This distinction becomes important when a prism is inserted into the beam. For example, a Dove prism can be used only in a truly parallel beam because the end faces are inclined to the lens axis. However, there are other types of prisms that can be used in a quasi-parallel beam (see Section V in Chapter 9).

V. LONGITUDINAL MAGNIFICATION

Longitudinal magnification is the ratio of the axial dimension of an image to the corresponding axial dimension of the object. These axial dimensions may be the physical sizes of object and image, or they may be a movement of object and image along the axis.

Two situations arise—the first is concerned with an infinitesimally small longitudinal dimension (enabling the magnification to be determined by differentiation), and the second is for a finite axial dimension.

A. A Very Small Axial Dimension

If the object and image have very small axial dimensions, we may write Δl for the axial dimension of the object and $\Delta l'$ for the corresponding axial dimension of the image. Then by differentiating Eq. (5) of Chapter 3 we find

$$-(n'/l'^2)\,\Delta l' = -(n/l^2)\,\Delta l,$$

from which the longitudinal magnification becomes

$$\overline{m} = \Delta l'/\Delta l = (n/n')(l'/l)^2.$$

Multiplying numerator and denominator by y^2 gives

$$\overline{m} = \Delta l'/\Delta l = nu^2/n'u'^2.$$

Therefore there is a *longitudinal magnification invariant* corresponding to the Lagrange invariant, namely, $\Delta l\,n\,u^2$. Since the ordinary

V. LONGITUDINAL MAGNIFICATION

magnification $m = nu/n'u'$, we see that

$$\overline{m} = (n'/n)m^2.$$

For a lens in air, the fact that $\overline{m} = m^2$ tells us that longitudinal magnification is always positive, so that if the object moves from left to right, the image must move from left to right also. On the other hand, for a mirror, the signs of n and n' are equal and opposite, so that when the object moves from left to right the image must move from right to left.

If the ordinary magnification is large, as in a microscope objective, the longitudinal magnification will be very large, which explains the small depth of field noticed in a microscope. On the other hand, in a camera the magnification is small, so the longitudinal magnification is very small, accounting for the great depth of field noticed in most cameras.

B. A Large Axial Dimension

If the longitudinal dimensions of object and image are large, we can still derive a useful expression for the longitudinal magnification. In Fig. 4.11 the magnifications of the objects at A and B are related to their image distances z'_A and z'_B from the rear focal point of the lens by

$$z'_A = -f'm_A \quad \text{and} \quad z'_B = -f'm_B.$$

When the object moves from A to B, the change in image distance $A'B'$ is therefore given by

$$A'B' = (z'_A - z'_B) = f'(m_A - m_B).$$

The corresponding change in the object distance AB is similarly

$$AB = f'\left(\frac{1}{m_B} - \frac{1}{m_A}\right).$$

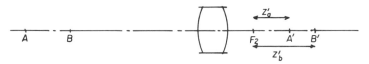

Fig. 4.11. Longitudinal magnification.

Therefore, the longitudinal magnification $A'B'/AB$ is given by

$$\overline{m} = \frac{f'(m_A - m_B)}{f'[(1/m_B) - (1/m_A)]} = m_A m_B.$$

This reduces to m^2 if the longitudinal dimensions AB and $A'B'$ are so small that the magnification does not change significantly when going from A to B.

VI. THE SCHEIMPFLUG CONDITION

Captain Theodor Scheimpflug of the Austrian army pointed out early in the present century that if the object plane is tilted relative to the lens axis, the image plane will also be tilted in such a way that the object plane, image plane, and median plane through the lens will all meet. This relation can easily be proved for the paraxial region by use of the longitudinal magnification formula. In Fig. 4.12, an object point A is imaged at A' and a neighboring point B on a sloping object is imaged at B'. The plane through AB and the plane through $A'B'$ are shown meeting the median plane through the lens at M and N, respectively. The middle of the lens is marked O. The heights OM and ON are given by

$$OM = -yl/z \quad \text{and} \quad ON = -y'l'/z',$$

where (y, z) are the coordinates of B relative to A and l the distance from A to the lens. In the image space we have similar quantities with

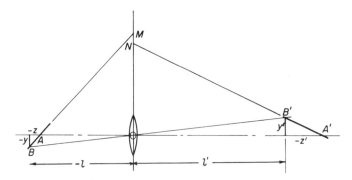

FIG. 4.12. The Scheimpflug condition.

primes. Since $y' = my$ and $l' = ml$ and since (by the longitudinal magnification rule) $z' = m^2 z$, we see that $OM = ON$, and this at once bears out Scheimpflug's rule. Although we have proved this only for the paraxial region, it actually holds well over a considerable field angle. This subject is discussed much more fully in Section IV of Chapter 15.

VII. FOCOMETRY

Focometry is the name given to the process of measuring the focal length of an actual lens. Two cases immediately arise: Are we to consider only the axial image point, or must we determine the location of the best average focal plane within the entire picture area? In the latter case, the measurement involves two parts—first the determination of the location of the best average focal plane, and then the determination of the focal length to that focal plane. Furthermore, it must be remembered that if the lens has distortion, the focal length will vary across the field.

This is a good place to point out that in most cases the focal length of a lens need not be known with any degree of precision. The ordinary camera owner could not care less what the actual focal length of his lens is; a rough value is quite close enough. Many lenses such as microscope objectives are not marked with the focal length at all. The only time the focal length of a camera lens must be known with high precision is with lenses used for photogrammetry i.e., the making of maps from aerial photographs.

A. Focal Length to an Axial Image

Several methods are available for measuring the focal length of a lens when the axial image is all that matters. They are as follows:

(*a*) *The Nodal Slide.* The nodal slide is a piece of apparatus in which the lens to be tested is mounted on a turntable over a vertical axis of rotation. The image formed by the lens of a distant object is observed through a low-power microscope, and when the turntable is rotated through a small angle, the image moves by the same amount that the N_2 point moves (Fig. 4.13). Thus, if the nodal point is displaced to one

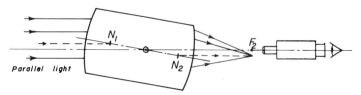

Fig. 4.13. The nodal slide.

side of the axis, the image will go out of focus, and if the nodal point is longitudinally displaced from the axis, the image will move sideways. With a little care it is usually easy to adjust the position of the lens on the turntable so that the image does not move or change in any way when the table is turned through a few degrees. The nodal point is now over the vertical rotation axis, and the image seen in the microscope is at the focal point, so the focal length is merely the distance from axis to image. There is usually no trouble in locating the nodal point, but the focal point requires some care. If the lens has any aberrations, the image seen in the microscope will not be perfect, and its position will vary with the lens aperture and the color of the light. Some workers prefer to use a single point source, and others use a resolution test chart or a page of print at the focus of a collimator lens.

(b) *By the Formula* $zz' = f^2$. This well-known "Newtonian" formula can be used to find the focal length of a medium-sized lens. The lens is mounted on a bench and faced toward a distant object or toward a collimator, and the position of the focal plane is noted. The lens is then turned around and the other focal plane is found. Finally, a convenient near object is used and its image found, so that the two distances z and z' can be determined, and the focal length found from the square root of the product.

(c) *The Two-Magnifications Method.* With a given pair of object and image positions, the magnification m_1 is found. The image is now moved through a distance d along the axis, and the object moved until a sharp image is obtained, with the new magnification being m_2. Then it is easily shown that

$$f = d/(m_1 - m_2).$$

This procedure is convenient for very small lenses such as microscope objectives because no measurements are made from the lens itself, but only of the magnifications at two different pairs of conjugates.

(*d*) *By A Foco-Collimator.* In this method, a collimator (coll) is equipped with a pair of parallel cross-wires in its focal plane, separated by a distance S. The test lens (t lens) is mounted in front of the collimator to form images of the two cross-wires in its focal plane, the separation of the images being s. Then, by simple proportion,

$$\frac{F_{\text{coll}}}{f_{\text{t lens}}} = \frac{S}{s}.$$

Often, the ratio of the focal length of the collimator to the separation of the cross-wires is made some whole number such as 100 or 1000, in which case the focal length of the test lens is that multiple of the separation of the images in its focal plane.

B. Focal Length for a Wide Angular Field

If the test lens covers a wide angular field and it is necessary to determine the focal length with high precision, the first task is to locate the plane of best average definition over the entire picture area. During World War II, when many types of lens were being used on aerial cameras, two useful terms were introduced, namely, AWAR and BADOPA.[1] The acronym AWAR stands for area weighted average resolution. To determine this, the picture area of the camera is divided into a number of circular zones, the resolving power of the lens is carefully measured in each zone, and the weighted average resolution is found. This is the AWAR for that particular focal plane, and its value varies from focal plane to focal plane along the axis. It is thus possible to determine the BADOPA plane, which is the plane of best average definition over the picture area, and it is this plane that must be used in the measurement of the focal length of the lens.

To determine the focal length, a film is exposed to a single swinging collimator or set of fixed collimators at a series of known obliquity angles ω at the lens (Fig. 4.14). By measuring the radial distances h' of the various images out from the lens axis, we can calculate the focal length at each obliquity by

$$f' = h'/\tan \omega.$$

---

[1] MIL-STD-150A, "Photographic lenses." Superintendent of Documents, U.S. Government Printing Office, Washington, D.C., 1959.

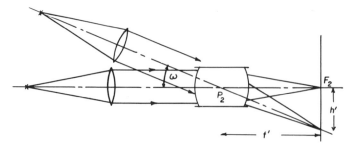

Fig. 4.14. The tilted collimator method of focometry.

A graph is now plotted connecting f' as ordinate against ω as abscissa, and the curve is extrapolated to the $\omega = 0$ ordinate to find the axial focal length. If the graph is a horizontal straight line, it indicates the absence of distortion, but if f' varies with ω, it provides a measure of the distortion across the field. The height of the best-fitting horizontal straight line through the graph gives what is known as the calibrated focal length in the presence of distortion.

VIII. AUTOFOCUS MECHANISMS

Various mechanisms have been proposed for use on enlargers or printers to maintain the image automatically in focus while the magnification is varied by moving the lens and the conjugate planes. The following are some of the more important mechanisms.

(a) *Two Sliding Rods.* In this arrangement, a pair of rods rigidly connected at right angles is pivoted at the elbow, which is located at a distance f to one side of the axis, f being the focal length of the lens. Two sliders ride on the rods, as shown in Fig. 4.15, one to drive the

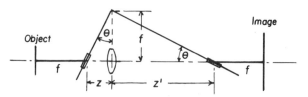

Fig. 4.15. The two-sliders autofocus mechanism.

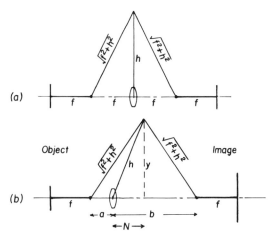

FIG. 4.16. The three linked rods autofocus mechanism.

object and the other to drive the image. The actual object and image are mounted at a distance f from the respective sliders. Then $z = f \tan \theta$ and $z' = f/\tan \theta$, from which $zz' = f^2$.

(b) *Three Linked Rods.* One very practical arrangement consists of three rods joined at the top by a pivot and connected at the bottom to the object, lens, and image, respectively, as shown in Fig. 4.16. The lengths of the rods are $\sqrt{f^2 + h^2}$, h, and $\sqrt{f^2 + h^2}$. In the mid position, the magnification is unity and the distance from object to image is $4f$. When the lens is moved through a distance N along the axis, the image plane moves away from its initial position, and the height Y of the pivot above the axis is given by the following three relations:

(a) $Y^2 = (f^2 + h^2) - (a + N)^2$,
(b) $Y^2 = h^2 - N^2$,
(c) $Y^2 = (f^2 + h^2) - (b - N)^2$.

From Eqs. (a) and (c) we find that $N = \frac{1}{2}(b - a)$, and from Eqs. (a) and (b) we obtain $N = (f^2 - a^2)/2a$. Equating these two expressions for N gives $f^2 = ab$, fully in accordance with the familiar formula $zz' = -f'^2$. Since h has dropped out of the formula, it may be assigned arbitrarily, and if a wide range of magnifications is desired, h should be made fairly large. Of course, some mechanical means must be provided to move the image plane and the lens at appropriate relative

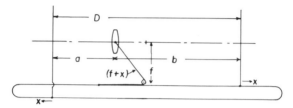

Fig. 4.17. The endless belt autofocus mechanism.

speeds to avoid jamming the mechanism.

(c) *The Endless Belt Method.* In this apparatus an endless belt or chain passing over a pair of pulleys is made to move the object and image in opposite directions (Fig. 4.17), with the lens being attached to an auxiliary belt passing over a small pulley situated a transverse distance from the lens equal to its focal length. When the chain is moved to change magnification, the lens belt is lengthened, and the lens is free to move in either direction along the axis.

In the mid position, at unit magnification, the object and image distances from the lens are each equal to $2f$. Now suppose that the chain is moved through a distance x as shown in Fig. 4.17. The lens belt will have increased in length from f to $f + x$, and the lens therefore moves along the axis by a distance $\sqrt{(f+x)^2 - f^2}$. Hence the object and image distances become

$$a = 2f + x - \sqrt{(f+x)^2 - f^2} \quad \text{and} \quad b = 2f + x + \sqrt{(f+x)^2 - f^2}.$$

The distances a and b are the conjugate distances of the lens, and on adding $1/a$ to $1/b$, we obtain $1/f$ in the usual way, thus demonstrating the validity of this procedure as an autofocus mechanism.

As an example, if $f = 10$ in., then for $m = 1$ we have $a = b = 20$ in. At $m = 2$, the separation between object and image increases from 40 to 45 in., so that the main chain has to move through 2.5 in. The lens belt then increases in length from 10 to 12.5 in., so that the lens moves along the axis through 7.5 in., making $a = 20 + 2.5 - 7.5 = 15$ in. and $b = 20 + 2.5 + 7.5 = 30$ in.

(d) *Cam Methods.* A large number of methods have been devised involving some sort of cam to keep the image in focus while the object and image, or the object and lens, are moved to change the magnification. One such arrangement is shown diagrammatically in Fig. 4.18. Here a bell-crank lever rests on a linear cam, to move the lens forward

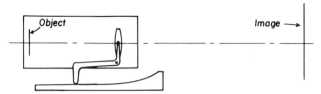

Fig. 4.18. A cam-driven autofocus mechanism.

at the right speed while the lens carrier and the object plane are moved along the axis. This arrangement has been used in an autofocus enlarger and also to project a variable-sized image for sight testing.

This is a good place to point out that if an enlarger is being used at or near unit magnification, it cannot be focused by moving the lens but only by moving either the object of image plane. Moving the lens in such a case merely changes the image size but not the image position.

CHAPTER 5

Multilens Systems

I. GRAPHICAL CONSTRUCTION OF AN IMAGE

If we know the location of an object in relation to a lens, there are three known rays that can be drawn, any two of which are sufficient to locate the image:

(a) a ray from the top of the object entering parallel to the axis, which passes through the F_2 point on the other side of the lens;
(b) a ray from the top of the object entering through the F_1 point and emerging parallel to the axis on the other side of the lens;
(c) a ray entering toward the first principal point P_1 and emerging at the same slope through the second principal point P_2. If the lens is thin, this ray passes straight through the middle of the lens, and it is the easiest ray to use in locating the image.

If the object is virtual, i.e., if it is formed by rays entering the lens and converging toward an object point, we do not need to know which rays were used to locate the object, but only the location of the object itself. If the lens is negative, remember that the F_2 point is on the left of the lens and the F_1 point on the right.

As an example, we may consider the system of two thin lenses shown in Fig. 5.1. Two known rays are drawn from the top of the object B through the first lens, one entering through F_A and the other entering through the middle of the lens. The intersection of the emerging portions of these rays locates the image of B at B', which now

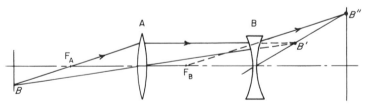

FIG. 5.1. Graphical construction of an image.

becomes the object for the second lens. One ray through the second lens is already available; it is the ray between the lenses parallel to the axis which emerges through the F_B point of the second lens. The other ray joins B' to the midpoint of lens B and passes on. These two rays cross at B'', which is the final image of B.

II. RAY TRACING THROUGH A SYSTEM OF SEPARATED THIN LENSES

We can readily adapt the familiar y-nu method of tracing a paraxial ray to the case of a succession of thin lenses separated by air, with each lens being defined by its power and diameter only. The formulas to be used are

$$u' = u - y\phi \quad \text{and} \quad y_2 = y_1 + (d'u')_1.$$

As before, y represents the height of incidence of the ray at each thin lens and u the slope of the ray between lenses. The lens powers are represented by ϕ and their separations by d. The calculation is performed in columns, one for each thin lens (see Table I). The lens system being traced is shown in Fig. 5.2.

TABLE I

Tracing a Ray Through a Series of Thin Lenses

f'		6.0		−3.3		5.1	
$-\phi$		−0.166667		0.30303		−0.19607	
d			1.34		1.34		
y		1.0		0.77666		0.86869	$l' = 8.546$ $f' = 9.838$
u	0.0		−0.16667		0.06868		−0.10164	

No refractive indices are entered and only the lens powers and separations are required. The five significant figures used here represents a reasonable practical value. The computation proceeds as in Section IVB Chapter 3, with each figure being the sum of the previous figure plus the product of the other dimension with the number immediately above it. The example given here assumes an object at infinity in order to calculate the focal length and the back focal distance. Of course, any object distance may be used, with the initial value of y being usually assigned as 1.0, and the initial u being given by $u_1 = -y_1/l_1$.

5. MULTILENS SYSTEMS

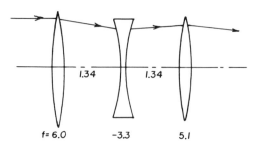

FIG. 5.2. Example of the $(y-u)$ ray-tracing method.

Although this procedure is intended for use with thin lenses spaced apart in air, it can be used with thick single lenses or a parallel plate of glass, with the thick lens being expressed as two thin lens elements separated by glass and the optical separation (for calculation purposes) being equal to the actual thickness divided by the refractive index. Thus a 50-mm plate of glass of index 1.62 is written as a pair of zero-power lenses separated by 30.8642 mm. This is particularly convenient if a reflection prism is inserted in a system, because, optically, a prism is nothing but a thick plate of glass with one or more plane mirrors incorporated into it. For a two-lens system we may perform the calculation in general algebraic terms instead of with numbers (see Table II).

TABLE II

TWO THIN LENSES SEPARATED BY AIR

$-\phi$ d		$-\phi_A$ d		$-\phi_B$	
y u	0	1.0 $-\phi_A$		$1 - d\phi_A$	$-\phi_A - \phi_B(1 - d\phi_A)$

For the two-lens system assumed in Table II, the focal length is given by

$$1/f' = -y_A/u'_B = \phi_A + \phi_B - d\phi_A \phi_B. \tag{1}$$

The back focal distance is then found as

$$l' = -y_B/u'_B = f'(1 - d\phi_A).$$

These formulas will often be found convenient.

III. TWO THIN LENSES WITH A REAL OBJECT AND IMAGE

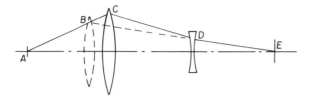

FIG. 5.3. A telephoto system at finite magnification.

III. TWO THIN LENSES WITH A REAL OBJECT AND REAL IMAGE

A situation which often arises is that in which a real object and real image lie in fixed planes and we wish to achieve a specified magnification between them.

The situation is indicated in Fig. 5.3. We draw incident and emerging rays from the axial points of object and image at such slopes as to yield the desired magnification m. Extending these rays until they cross will locate a single lens that will do the job, indicated at B. However, if there is some reason why this single lens cannot be used, it will be necessary to use two lenses as indicated at C and D in Fig. 5.3. The entering and emerging rays are extended until they intersect the thin lenses, and all we have to do is to select suitable lens powers to make the intermediate ray join the points C and D. There is an infinite number of possible solutions.

If the lens locations are specified, we can operate the $y-u$ table backward, starting with the ray data and working upward to fill in the lens powers. In Table III the given data are stated in full and the unknown quantities are shown in parentheses. The standard relations

TABLE III

SOLUTION FOR LENS POWERS

	$-\phi$		$(-\phi_A)$		$(-\phi_B)$		
	d			j			
$l=-i$	y		mi		k		$l'=k$
	u	m		(u)		-1	

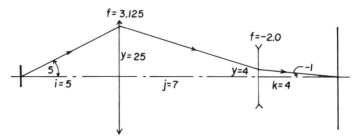

FIG. 5.4. A magnification of 5.0 obtained by two lenses.

are

$$u = (k - mi)/j, \quad \phi_A = (m - u)/mi, \quad \text{and} \quad \phi_B = (u + 1)/k.$$

Example. Suppose we are given $i = 5, j = 7, k = 4$, and $m = 5$. Then $y = mi = 25$ and $u = (4 - 25)/7 = -3$; $\phi_A = (5 + 3)/25 = 0.32$; and $\phi_B = (1 - 3)/4 = -0.5$. The final situation is drawn to scale in Fig. 5.4. Generally the known data are entered directly into the table in numerical form and the numbers are determined by ordinary arithmetic (rather than by algebraic formulas).

A much more difficult problem arises if we know the lens powers but not their locations (see Table IV). The standard relations are now

$$m - \phi_A mi = u, \qquad (2a)$$
$$mi + ju = k, \qquad (2b)$$
$$u - k\phi_B = -1, \qquad (2c)$$
$$i + j + k = D, \qquad (2d)$$

where D is the distance from object to image. These four equations must be solved simultaneously for i, j, k, and u. The solution may be performed in this way: From (2b) we obtain $u = (k - mi)/j$, and

TABLE IV

SOLUTION FOR LENS POSITIONS

	$-\phi$ d	$-\phi_A$ (j)	$-\phi_B$	
$l = -i$	y u	(mi) m (u)	(k) -1	$l' = k$

substituting this into (2a) gives

$$k = -jm + mi - mij\phi_A. \tag{2e}$$

Substituting u into (2c) gives

$$k = \frac{mi - j}{1 - j\phi_B}, \tag{2f}$$

and lastly from (2d) we obtain

$$k = D - i - j. \tag{2g}$$

From (2e) and (2f) we find, after some reduction, that

$$i = \frac{1 + m - jm\phi_B}{m\phi_A + m\phi_B - mj\phi_A\phi_B}, \tag{2h}$$

whereas from (2e) and (2g) we obtain

$$i = \frac{j(1 + m) - D}{mj\phi_A - (1 + m)}. \tag{2i}$$

Equating (2h) and (2i) gives us a quadratic for j, namely,

$$j^2 - jD + \left(\frac{1}{\phi_A} + \frac{1}{\phi_B}\right) - \frac{(m + 1)^2}{m\phi_A\phi_B} = 0 \tag{2j}$$

Having solved this for j, we obtain i by (2h) or (2i), and then k from (2g).

Example. From the previous example we found that $m = 5, D = 16$, and $\phi_A = 0.32$ and $\phi_B = -0.5$. Inserting these values into (2j) gives $j = 9$ or 7. From (2i) we find that $i = 4.5238$ or 5.0, and then from (2d) we obtain $k = 2.4762$ or 4.0. The second solution is the one we used previously, whereas the first solution is new.

A. Telephoto and Reversed Telephoto Systems

Telephoto and reversed telephoto systems are used extensively in photography, and each consists of a widely separated positive and negative component (see Fig. 5.5). It will be seen that in the telephoto system, with the positive lens in front, the total length from front

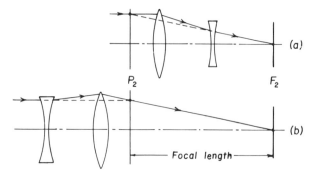

Fig. 5.5. (a) Telephoto and (b) reversed telephoto systems.

vertex to focal plane is less than the focal length, with the ratio of total length to focal length being the *telephoto ratio,* generally about 0.8 or a little over.

In the reversed telephoto, on the other hand, the negative component is in front, and the back focal distance is longer than the focal length, a very desirable feature in wide-angle lenses for SLR cameras in which a clearance of about 35 mm behind the lens is necessary to clear the rocking mirror.

B. A Bravais System

A *Bravais system* is a lens or combination of lenses such that the object and image lie in the same plane, although one is in the object space and the other in the image space. The simplest Bravais system is a hemisphere of glass, with object and image lying on the flat side of the hemisphere. For rays entering from the left, as shown in Fig. 5.6a, the magnification is $1/n$.

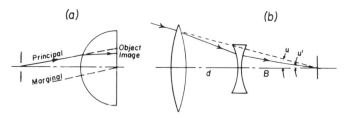

Fig. 5.6. Two Bravais systems.

III. TWO THIN LENSES WITH A REAL OBJECT AND IMAGE

TABLE V
A Bravais System

	$-\phi$ d	$-\phi_A$ d	$-\phi_B$	
$l = d + B$	y u	$-m$	$\dfrac{m(d+B)}{}$ $\dfrac{B - m(d+B)}{d}$	B -1	$l' = B$

Other Bravais systems can be constructed from two lenses spaced apart. Such a system is shown in Fig. 5.6b, with the magnification being given by the ratio of the entering ray slope u to the emerging slope u' (see Table V). Hence

$$\phi_A = \frac{B(m-1)}{md(d+B)} \quad \text{and} \quad \phi_B = \frac{(1-m)(d+B)}{dB}.$$

Example. As an example, suppose $d = 2$, $B = 2$, and $m = 3$. Then $\phi_A = \tfrac{1}{6}$ and $\phi_B = -2.0$.

Bravais systems of this kind are sometimes made with cylindrical surfaces. These give an anamorphic magnification in one meridian only, and they are intended to be used in front of a camera or enlarger lens. It is obvious that the unmagnified image in the null meridian and the magnified image in the power meridian must be made to fall into the same plane, and hence a Bravais system of cylinders is necessary.

Ordinary Bravais systems made with spherical surfaces have been proposed for use in a film printer to take care of the occasional need for some unusual magnification not provided by the lenses normally available for the printer.

C. Afocal Systems

An afocal system is one in which parallel light both enters and emerges from the system. The focal and principal points are, therefore, all at infinity, and the focal length is infinite.

Although such a system cannot produce a real image of a distant

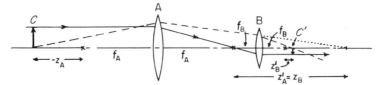

FIG. 5.7. An afocal system with a near object.

object, it certainly can form useful images of near objects. In Fig. 5.7, an afocal system consisting of two positive lenses separated by the sum of their focal lengths is shown with a near object at C. A ray entering parallel to the axis from the top of the object emerges parallel to the axis and necessarily passes through the top of the image at C'. Thus, wherever the image may be located, its magnification must be equal to the ratio of the focal length of the second lens to that of the first, or $m = f'_B/f'_A$.

Now suppose that the distance of the object from the anterior focus of lens A is $-z_A$. The distance of the intermediate image from the posterior focus of lens A is therefore $z'_A = -f'^2_A/z_A$. Because the adjacent foci of the two lenses coincide, z_B of the second lens is equal to z'_A of the first lens, that is, $-f'^2_A/z_A$. Consequently, the z'_B of the second lens will be given by $-f'_B{}^2/z_B = (f'_B/f'_A)^2 z_A$.

It is worth noting that because the magnification is constant for all object distances, the longitudinal magnification is also constant. Therefore, if the object distance changes by an amount D, the image distance changes by an amount $m^2 D = (f'_B/f'_A)^2 D$.

The most interesting special case of an afocal system is one in which the two lenses have the same focal length, and the magnification for all object distances and longitudinal magnifications is therefore unity. If such a system is used with a near object, the position and size of the image remain constant while the optical system is moved along the axis. The separation between object and image is fixed at four times the focal length of the lenses.

It will be seen later that a simple telescope is an afocal system.

D. A Two-Lens Zoom System

A number of wide-angle zoom lenses have recently appeared with a negative lens in front, a positive lens behind, and with both lenses

III. TWO THIN LENSES WITH A REAL OBJECT AND IMAGE

being moved to vary the focal length and hold the image in a fixed position. The range of focal lengths in these systems is usually small, say about 2:1, but they can be made to cover exceptionally wide angular fields. They are much more compact than other types of zoom systems.

It should be noted that, in any two-lens system with a variable lens separation, the back focus varies linearly with the focal length. This is easily seen by expressing l' and f' as functions of the separation d and then eliminating d between the two expressions. Thus we have

$$1/f' = \phi_A + \phi_B(1 - d\phi_A)$$

and

$$l' = f'(1 - d\phi_A).$$

Equating the two values of $1 - d\phi_A$ gives

$$l' = f_B - (f_B/f_A)f'.$$

Example. As an example we may quote from a recent patent in which the focal lengths of the lens components were -33 mm and $+29$ mm, respectively, and the separation varied from 36 to 23 mm between adjacent principal planes, although of course the real lenses contained several elements and were quite thick (see Table VI). The distances given in Table VI are plotted in Fig. 5.8. A similar system will be shown in Fig. 15.11. It is interesting to note that the "total length"

TABLE VI

A Two-Component Zoom System[a]

Focal length	Separation	Back focus	Total length
f'	d	l'	$d + l'$
24	35.875	50.091	85.966
26	32.808	51.848	84.656
28	30.179	53.606	83.785
30	27.900	55.364	83.264
32	25.906	57.121	83.027
34	24.147	58.879	83.026
35	23.345	59.758	83.103

[a] All dimensions are in millimeters.

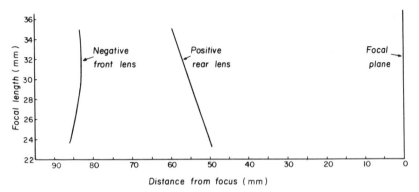

FIG. 5.8. Movements of a two-component zoom.

from the front element to the focal plane drops to a minimum and rises again, with the minimum value of the total length occurring when the overall focal length is equal to the absolute value of the focal length of the front element, namely, 33 mm. This property of a two-component system can be readily proved. The total length L is equal to $l' + d$, hence

$$L = l' + d = f_B - (\phi_A/\phi_B)f' + d.$$

Differentiating with respect to d gives

$$\frac{\partial L}{\partial d} = 1 - \left(\frac{\phi_A}{\phi_B}\right)\left(\frac{\partial f'}{\partial d}\right).$$

But by Eq. (1) we see that

$$\frac{\partial f'}{\partial d} = \frac{\phi_A \phi_B}{(\phi_A + \phi_B - d\phi_A\phi_B)^2} = \phi_A\phi_B f'^2.$$

Hence, for the minimum value of L, we have $\partial L/\partial d = 1 - f'^2/f_A^2 = 0$, from which $f'^2 = f_A^2$. The overall focal length is, therefore, equal to the absolute value of f_A. It should be noted that if the front element is negative, as in this example, it is the minimum overall focal length that is significant. In the reverse case when the front lens is positive and the rear lens negative, it is the maximum focal length that is equal to the focal length of the front lens. This occurs when the negative rear lens is actually at the focal plane, and $d = f_A$.

IV. THE MATRIX APPROACH TO PARAXIAL RAYS[1,2]

It has been pointed out by Gauss and others that the similarity between the paraxial equations for nu and y suggests a simple matrix formulation for these relations.

The rules of matrix algebra are simple. Suppose we have two simultaneous equations in x and y such as

$$A = ax + by,$$
$$B = cx + dy.$$

Then in matrix notation we can write

$$\begin{bmatrix} A \\ B \end{bmatrix} = \begin{bmatrix} a & b \\ c & d \end{bmatrix} \begin{bmatrix} x \\ y \end{bmatrix}.$$

Furthermore, the product of two matrices is another matrix, of which the elements are

$$\begin{bmatrix} a & c \\ b & d \end{bmatrix} \begin{bmatrix} e & f \\ g & h \end{bmatrix} = \begin{bmatrix} ae + cg & af + ch \\ be + dg & bf + dh \end{bmatrix}.$$

To apply matrix notation to the case of a paraxial ray through a lens, we note that, for the first lens surface,

$$(nu)'_1 = (nu)_1 - y_1 \phi_1,$$
$$y_1 = y_1.$$

In matrix notation these formulas become

$$\begin{bmatrix} (nu)'_1 \\ y_1 \end{bmatrix} = \begin{bmatrix} 1 & -\phi_1 \\ 0 & 1 \end{bmatrix} \begin{bmatrix} (nu)_1 \\ y_1 \end{bmatrix}. \tag{3}$$

This square matrix is known as the *refraction matrix* for the first surface. The transfer to the next surface is performed by

$$(nu)_2 = (nu)'_1 \quad \text{and} \quad y_2 = y_1 + (nu)'_1(t/n)'_1,$$

---

[1] W. Brower, "Matrix Methods in Optical Instrument Design." Benjamin, New York, 1964.

[2] H. Kogelnik, "Paraxial ray propagation," in "Applied Optics and Optical Engineering" (R. R. Shannon and J. C. Wyant, eds.), Vol. 7, p. 156. Academic Press, New York, 1979.

5. MULTILENS SYSTEMS

which in matrix notation becomes

$$\begin{bmatrix} (nu)_2 \\ y_2 \end{bmatrix} = \begin{bmatrix} 1 & 0 \\ (t/n)'_1 & 1 \end{bmatrix} \begin{bmatrix} (nu)'_1 \\ y_1 \end{bmatrix}.$$

This square matrix is known as the *transfer matrix* from surface 1 to surface 2. But the last matrix here is the left side of Eq. (3). Substituting this into the last relation gives

$$\begin{bmatrix} (nu)_2 \\ y_2 \end{bmatrix} = \begin{bmatrix} 1 & 0 \\ (t/n)'_1 & 1 \end{bmatrix} \begin{bmatrix} 1 & -\phi \\ 0 & 1 \end{bmatrix} \begin{bmatrix} (nu)_1 \\ y_1 \end{bmatrix}.$$

We can verify this by multiplying the two square matrices together. This gives

$$\begin{bmatrix} (nu)_2 \\ y_2 \end{bmatrix} = \begin{bmatrix} 1 & -\phi_1 \\ (t/n)'_1 & 1 - \phi_1(t/n)'_1 \end{bmatrix} \begin{bmatrix} (nu)_1 \\ y_1 \end{bmatrix},$$

which correctly represents the two equations

$$(nu)_2 = (nu)_1 - y_1\phi_1,$$
$$y_2 = y_1 + [(nu)_1 - y_1\phi_1](t/n)'_1.$$

We can extend this argument to a system containing any number k of surfaces, giving

$$\begin{bmatrix} (nu)'_k \\ y_k \end{bmatrix} = \underbrace{\begin{bmatrix} 1 & -\phi_k \\ 0 & 1 \end{bmatrix}}_{\substack{\text{refraction} \\ \text{at} \\ \text{surface } k}} \underbrace{\begin{bmatrix} 1 & 0 \\ (t/n)'_{k-1} & 1 \end{bmatrix}}_{\substack{\text{transfer} \\ \text{from} \\ (k-1) \text{ to } k}} \underbrace{\begin{bmatrix} 1 & -\phi_{k-1} \\ 0 & 1 \end{bmatrix}}_{\substack{\text{refraction} \\ \text{at} \\ \text{surface } k-1}} \cdots$$

$$\underbrace{\begin{bmatrix} 1 & 0 \\ (t/n)'_1 & 1 \end{bmatrix}}_{\substack{\text{transfer} \\ \text{from} \\ 1 \text{ to } 2}} \underbrace{\begin{bmatrix} 1 & -\phi_1 \\ 0 & 1 \end{bmatrix}}_{\substack{\text{refraction} \\ \text{at} \\ \text{surface } 1}} \begin{bmatrix} (nu)_1 \\ y_1 \end{bmatrix}.$$

The product of all the square matrices, taken in order, is another square matrix which is a pure property of the lens. It can be written as

$$\begin{bmatrix} B & -A \\ -D & C \end{bmatrix},$$

IV. THE MATRIX APPROACH TO PARAXIAL RAYS

with the property that the determinant of this matrix, $BC - AD$, equals 1.0. The four quantities A, B, C, and D are called the Gauss constants of the lens. Here A is the lens power, B the ratio of the front focal distance to the front focal length, and C the ratio of the back focal distance to the rear focal length. We find D by $(BC - 1)/A$. Knowing the four elements of this matrix, we can immediately find the values of $(nu)'_k$ and y_k for any ray defined by its entering values of $(nu)_1$ and y_1.

As an example we will take the doublet in Section IVB of Chapter 3. We find that for this lens the Gauss constants are

$$A = 0.0833332 \quad = 1/f',$$
$$B = 0.9800774 \quad = -FF/f',$$
$$C = 0.9404865 \quad = BF/f',$$
$$D = -0.9390067 = (BC - 1)/A.$$

Using this matrix, if $(nu)_1 = 0.02$ and $y_1 = 1.0$, for example, then we find that $(nu)'_3 = -0.063732$ and $y_3 = 0.959267$, both agreeing perfectly with the results of a direct paraxial ray trace.

In practice it is generally easiest to find the lens power and the positions of the focal points by tracing right-to-left and left-to-right paraxial rays through the lens and then determine the Gauss constants by their meanings given above. Then for any ray defined by its $(nu)_1$ and y_1 we have for the emerging ray

$$(nu)'_k = B(nu)_1 - Ay_1,$$
$$y_k = -D(nu)_1 + Cy_1.$$

A. A Single Thick Lens

Since $\phi_1 = (N-1)/r_1$ and $\phi_2 = (1-N)/r_2$, we see that the Gauss constants of a single thick lens are:

$$A = \phi_1 + \phi_2 - (t/N)\phi_1\phi_2,$$
$$B = 1 - (t/N)\phi_2,$$
$$C = 1 - (t/N)\phi_1,$$
$$D = -(t/N).$$

Example. Suppose $r_1 = 5.0$ and $r_2 = -10.0$ for a biconvex lens, with $t = 1.5$ and $N = 1.523$. Then $(N - 1)/N = 0.343401$, hence $1/f' =$

0.151512. This gives

$$f' = 6.600137,$$
$$FF = -6.260163,$$
$$BF = 5.920189,$$

Hence

$$l_{pp} = 0.339974,$$
$$l'_{pp} = -0.679948,$$

and the Gauss constants are

$$A = 0.151512,$$
$$B = 0.948490,$$
$$C = 0.896980,$$
$$D = -0.984898,$$

with $BC - AD = 1.0$ (as it should).

B. A Succession of Separated Thin Lenses

If we apply the matrix notation to a succession of thin lenses separated by air, the refraction matrix becomes $\begin{bmatrix} 1 & -\phi \\ 0 & 1 \end{bmatrix}$ for each thin lens, and the transfer matrix becomes $\begin{bmatrix} 1 & 0 \\ d & 1 \end{bmatrix}$ for each space between lenses. Thus, for a system of two thin lenses, we have

$$\begin{bmatrix} u'_B \\ y_B \end{bmatrix} = \begin{bmatrix} 1 & -\phi_B \\ 0 & 1 \end{bmatrix} \begin{bmatrix} 1 & 0 \\ d & 1 \end{bmatrix} \begin{bmatrix} 1 & -\phi_A \\ 0 & 1 \end{bmatrix} \begin{bmatrix} u_1 \\ y_1 \end{bmatrix}.$$

Since the product of these three matrices must be equal to $\begin{bmatrix} B & -A \\ -D & C \end{bmatrix}$, for two thin lenses we have

$$A = \phi_A + \phi_B - d\phi_A\phi_B,$$
$$B = 1 - d\phi_B,$$
$$C = 1 - d\phi_A,$$
$$D = -d.$$

V. CYLINDRICAL LENSES

A cylindrical lens has a cylindric surface on one side and usually a plane surface on the other side, although biconvex and biconcave

V. CYLINDRICAL LENSES

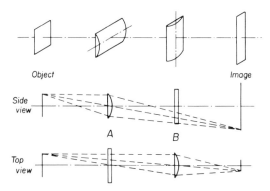

Fig. 5.9. Anamorphic imagery with two cylinder lenses.

cylindrical lenses have been made. In this case it is essential that the cylinder axes on the two sides be strictly parallel, for the slightest rotation of one axis relative to the other precludes the possibility of good imagery.

A cylindrical lens has a *power* meridian and a *null* meridian, which are perpendicular to each other. Looking at a cylinder lens end-on gives the appearance of a plano convex lens, while looking at it from the side gives the impression of a plate of glass.

A pair of cylindrical lenses with their axes at right angles has been used to distort an image anamorphically, as shown in Fig. 5.9. An oblique view of the system is shown at the top, with plan and elevation views below. In the side view, only the first lens has any power, whereas in the plan view only the second lens is operative. As a result, a square object is imaged as a tall rectangle (the height being greater than the height of the object), whereas its width is less than the width of the object.

Another application of cylindrical lenses to compress or stretch an image anamorphically has been employed in the CinemaScope system of motion picture photography and projection, shown schematically in Fig. 5.10. A reversed Galilean telescope, made entirely of cylindrical surfaces with all their cylinder axes vertical, is mounted in front of an ordinary movie camera lens. The attachment compresses the image in its power meridian without any effect in the null meridian, thus a square object is imaged as a narrow rectangle in the horizontal direction with no change in the vertical meridian. A similar anamorphoser

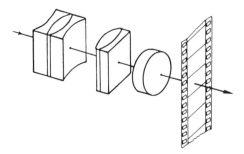

FIG. 5.10. The CinemaScope system.

is mounted in front of the projection lens which stretches the image horizontally to restore its original shape on the screen.

The image of a point of light P, formed by a cylindrical lens, is a thin line of light parallel to the cylinder axis, as indicated by I in Fig. 5.11. The length of the line image is determined by the length of the cylinder lens. A line source in the same direction as the cylinder axis is also imaged as a line, but now the ends of the image gradually fade out instead of being cut off abruptly.

A cylinder lens has been used to construct a sine-wave bar chart for MTF measurement. The source of light is now a suitably shaped area, and its image is a reasonably uniform bar of light parallel to the cylinder axis. The cross section of flux in the image reproduces the cross section of area in the object, measured in a direction perpendicular to the cylinder axis. Hence, to produce a sine-wave distribution of light in the image requires an object shaped like a sine wave. However, because of the well-known nonlinearity of the photographic process, a photographic record of the image will not in general have the desired sine-wave distribution of luminance, and it is necessary to make a series of careful distortions of the object area until the desired result is obtained. This is far from being a simple procedure.

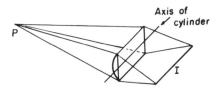

FIG. 5.11. Formation of a line image from a point source.

A. Crossed Cylinders

If two thin cylindrical lenses having powers ϕ_A and ϕ_B, where $\phi_A > \phi_B$, are crossed perpendicularly, the result is the combination of a sphere with power ϕ_B and a cylinder of power $\phi_A - \phi_B$, with the axis of the equivalent cylinder being the same as the axis of the original cylinder ϕ_A. If the cylinders are crossed at some angle other than 90°, the result is still equivalent to a sphere plus a cylinder, but the whole calculation is now quite complicated.[3]

---

[3] L. C. Martin, "Applied Optics," Vol. 1, p. 288. Pitman, London, 1930.

CHAPTER 6

Oblique Beams

I. MERIDIONAL RAYS

A. Axial Rays

So far we have considered only axial beams through a lens, i.e., beams of light that are symmetrical about the lens axis, with the object point being on the axis and either in front of the lens (real) or behind it (virtual).

If the object is at infinity, the entering slope angle of the marginal ray is zero and its height of incidence is given by the focal length of the lens divided by twice the *F*-number. Hence

$$U_1 = 0 \quad \text{and} \quad Y_1 = f'/2N,$$

where N is the *F*-number of the lens.

If the object is at a finite distance, the extreme marginal ray is defined by its entering slope angle and its initial Q value:

$$\sin U_1 = \frac{1}{\text{twice the effective } F\text{-number}} = \frac{1}{2N(1 + 1/m)},$$

and

$$Q_1 = -L \sin U_1.$$

Here N is the infinity *F*-number and m the image magnification (considered positive for a real image). We are assuming that the pupil magnification is unity. Notice that these starting data for a near object do not depend on the focal length of the lens; only the object distance, *F*-number, and magnification are significant.

B. Oblique Rays

We must now consider an oblique beam passing through a lens from an object point lying to one side of the axis, a so-called extraaxial object point. We know where the object is located, but we need further information to determine the highest and lowest oblique rays that will pass through the system. In many lenses there is an aperture stop (often an iris diaphragm) somewhere in the lens that limits the size of oblique beams passing through it. Once the location of the aperture stop is known, its diameter can be found from the extreme axial ray passing through the system.

C. Vignetting

At small obliquity, the highest and lowest rays passing through the lens are usually those that strike the top and bottom of the aperture stop. However, at greater obliquities, the extreme upper and lower rays are often limited by the front and rear lens openings, respectively (see Fig. 6.1). The failure of an oblique beam to fill the aperture stop is known as *vignetting,* and it leads to a lowering of the image illumination at increasing distances out from the axis. In some lenses, the front and rear openings have been deliberately enlarged to reduce the vignetting, but it is customary in many lenses to make these openings

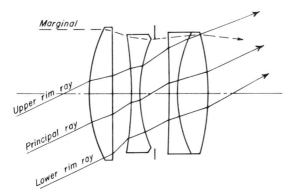

Fig. 6.1. Vignetting.

just large enough to pass the extreme marginal ray, and then vignetting begins at very small obliquities.

D. Principal and Chief Rays

The ray that passes through the middle of the aperture stop from an extraaxial object point is called the *principal ray* of an oblique beam. The ray that enters the lens midway between the highest and lowest rays of an oblique beam is called the *chief ray* of the beam. Obviously, in the absence of vignetting, the principal ray and the chief ray are identical, and the most desirable position for the iris in any lens is where the chief ray crosses the axis inside the lens, as indicated in Fig. 6.1. Unfortunately, it often happens in photographic objectives that, because of the particular arrangement of lens elements and airspaces, this is not always possible, and in such a case the principal ray and the chief ray may be quite different. When the iris diaphragm is stopped down to a small aperture, only the principal ray remains.

II. THE IRIS AND PUPILS OF A LENS SYSTEM

By analogy with the human eye, the limiting aperture that determines how large a beam shall pass through a lens is called the *iris,* and the image of the iris as seen from the object space is called the *pupil* of the lens. Actually, in any real lens there are two pupils—the *entrance* pupil being the image of the iris in the object space and the *exit* pupil being its image in the image space. The iris and two pupils of a lens are therefore all images of each other. Remember that these are paraxial images and they exhibit no vignetting; hence a paraxial principal ray enters a system through the middle of the entrance pupil, passes through the middle of the aperture stop (iris), and leaves through the middle of the exit pupil.

At higher obliquities, the pupils are liable to move longitudinally, laterally, and even become tilted by an amount depending on the obliquity, and it then becomes useless to speak of the pupils at all. However, in all cases the highest and lowest rays of an oblique beam

II. THE IRIS AND PUPILS OF A LENS SYSTEM

are limited by the aperture stop and often by the front and rear openings of the lens as well.

A. Entrance Pupil of a Fish-Eye Lens

A fish-eye lens is one capable of imaging an entire hemisphere in the object space onto a finite piece of film by the deliberate introduction of a huge amount of negative distortion. In fact, the law connecting the image height h' with the angular subtense ω of the object is $h' = f\omega$, rather than the usual $h' = f' \tan \omega$.

Obviously, if the entrance pupil of a fish-eye lens were in a fixed plane perpendicular to the axis, no light whatever would enter the lens at 90° to the axis. In fact, in such a lens, the entrance pupil moves forward, sideways, and tilts over toward the incoming light as the obliquity is increased (see Fig. 6.2). This diagram was plotted accu-

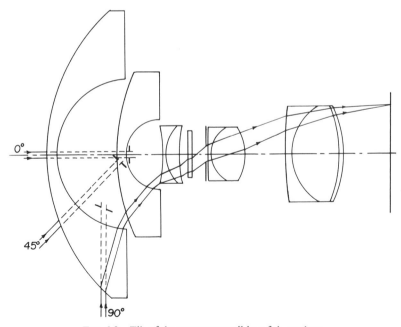

FIG. 6.2. Tilt of the entrance pupil in a fish-eye lens.

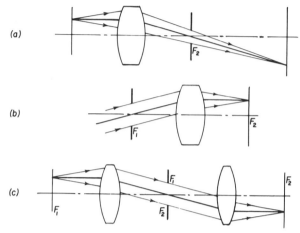

FIG. 6.3. Three telecentric systems.

rately for the lens shown in Miyamoto's article[1] in the *Journal of the Optical Society of America,* and the location of the entrance pupil, which is a virtual image of the aperture stop, is shown at obliquities of 0°, 45°, and 90°. Many wide-angle lenses of the reversed telephoto type have a large negative meniscus element in front, and they show the same phenomenon, as can be seen by looking into the front of the lens and tilting it to a considerable obliquity.

B. A Telecentric System

In most cases the aperture stop is located somewhere inside the lens. However, there are exceptions, as in eyepieces, condensers, and lenses used with a rocking mirror for scanning. An interesting special case is the telecentric system, in which one or both of the pupils are located at infinity and the principal ray enters or leaves the system parallel to the axis. Three examples of telecentric systems are shown in Fig. 6.3. System (a) is telecentric in the object space. The aperture stop is at the rear focal plane F_2 and the entering principal ray is parallel with the axis. System (b) is telecentric in the image space; the stop is at the front focus F_1 and the entering beam is shown coming from infinity. System (c) is telecentric at both ends, and the principal ray enters and leaves

[1] K. Miyamoto, "Fish-eye lens." *J. Opt. Soc. Am.* **54,** 1060 (1964).

parallel to the axis. Such a system is of course afocal, with the terms "telecentric" and "afocal" being applied, respectively, to systems in which either the principal rays or the marginal rays are parallel to the axis. A telecentric system is used wherever it is important for the beam as a whole to be perpendicular to an object or an image plane, e.g., in a contour projector or an orthographic camera system.

III. PARAXIAL TRACING OF AN OBLIQUE BEAM

Much preliminary information about a system, such as the required diameters of the various components, can often be obtained by tracing selected paraxial rays. For instance, tracing a 'marginal' paraxial ray by the $y-nu$ method gives us the height y at which the marginal ray strikes each surface or each lens element in the system. Of course, the ray must be made to enter the system at the correct values of y_1 and $(nu)_1$, corresponding to the desired Y_1 and $\sin U_1$ of the true marginal ray.

Next we trace a paraxial principal ray from an extraaxial object point through the center of the stop. To do this, it is first necessary to locate the entrance pupil by tracing a paraxial ray right-to-left from the center of the stop out into the object space. The desired paraxial principal ray is then directed toward the center of the entrance pupil from the specified object point. If the object is at infinity, the slope u_{pr} of the paraxial principal ray must be set equal to the tangent of the true obliquity angle, so that the ideal image height will be found as $h' = f'u_{pr}$.

Having traced these two paraxial rays, we can determine the y values of the upper (u) and lower (l) limiting oblique rays by

$$y_u = y_{pr} + y_{mar}$$

and

$$y_l = y_{pr} - y_{mar},$$

at each surface or each thin element, where pr is the principal and mar the marginal values. This saves a great deal of computing and enables us to make a rough scale drawing of the system for further study (see Fig. 6.4). Of course, as these are only paraxial rays, the drawing cannot be exact, but to make it exact would require the tracing of a large number of rays by accurate methods, including the decision as to how much vignetting should be permitted.

6. OBLIQUE BEAMS

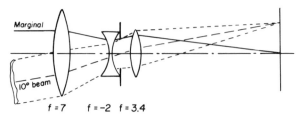

FIG. 6.4. An oblique paraxial beam through a system of thin lenses.

Example. As an example we shall consider an imaginary system consisting of three separated thin elements with an aperture stop, as shown in Fig. 6.4. The ray-tracing data are shown in Table I:

TABLE I

AN OBLIQUE PARAXIAL BEAM THROUGH A LENS

	f'	7	-2	(stop)	3.4	
	$-\phi$	-0.142857	0.5	0	-0.294118	
	d		3.0	0.5	1.1	
				Paraxial f/4 ray:		
	y	1.35227	.772726	0.869317	1.081817	$l' = 8.65454$
	u	0	-0.193181	0.193181	0.193181	$-0.125f' = 10.81816$
				Right-to-left stop ray:		
	y	-0.425	-0.05	0		
	u	.064286	0.125	0.10	$l_{pr} = -6.6111$	
				Paraxial principal ray at 10°		
$-l_{pr}u_{pr} = y_{pr}$		-1.165716	-0.137143	0	0.301716	$h' = 1.907533$
$\tan 10° = u_{pr}$		0.176327	0.342858	0.274287	0.274287	0.185547
				Upper rim ray		
$y_{pr} + y$		0.186	0.636	0.869	1.384	
				Lower rim ray		
$y_{pr} - y$		-2.518	-0.909	-0.869	-0.780	

(The upper and lower rim rays are shown in position.) We must now calculate the required F-numbers of the lens elements to see if it is possible to construct the system. The results are listed in the following tabulation.

Lens	Minimum aperture	Focal length	F-number
1	5.04	7.0	1.39
2	1.82	-2.0	1.10
3	2.77	3.4	1.23

It is clear from this analysis that all the lens elements, and especially the middle lens, must have an impossibly large aperture to avoid vignetting. For a practical system, we could reduce either the relative aperture or the angular field, or we could deliberately introduce some vignetting. The latter possibility is the most practical.

CHAPTER 7

The Photometry of Optical Systems

I. INTRODUCTION

With any optical instrument, a quantitative knowledge of the flow of light from object to image is of the greatest importance. It enables us to determine the illumination falling on the film in a camera, for instance, or the brightness of a star seen in a telescope, or an image projected on a screen. We live in a world of light, and we can see surrounding objects only if they have sufficient brightness.

We cannot see "light"; if we could, the entire night sky would be bright, as a large amount of light from the sun is streaming past the earth all night. We can see only the source from which the light comes. Thus our response to light is entirely different from our response to sound. We can be in a room full of sound and have no idea where it originates. Our ears can distinguish high notes from low notes, and even different types of sound, whereas our eyes can detect only the resultant mixture of wavelengths which we refer to as the "color" of the object.

A. RADIOMETRY AND PHOTOMETRY

It is essential to distinguish carefully between the basic concepts of radiometry and photometry. *Photometry* is concerned only with the integrated effect of all the wavelengths to which the eye is sensitive, i.e., from about 0.4 to about 0.75 μm. The basic unit of photometry is that of brightness, or more strictly *luminance,* because the eye responds only to brightness. On the other hand, *radiometry* is concerned with the emission and detection of radiant energy, taken wavelength by wavelength through the entire radiation spectrum from the extreme ultraviolet to the extreme infrared, with the detectors being generally responsive to radiant power, which is expressed in watts. In this chapter we shall consider only photometric concepts, assume that the light is "white," and leave problems of color aside for separate consideration.

II. PHOTOMETRIC DEFINITIONS

There are four basic concepts in photometry that must be clearly differentiated and understood. Two of these relate to the emission of light, one to the flow of light from a source to a receiver, and the last to the effect of light on a receiving surface.

A. Luminance

The concept of luminance (B), formerly called brightness, is basic to all photometry because the eye responds only to luminance. An object having a high luminance looks bright to our eyes, and conversely, an object having a low luminance looks dim to our eyes. The apparent brightness of an object depends on three factors—the luminance of the object, the size of the pupils of our eyes, and our state of adaptation. If we have been in a dark room for some time, our eyes become dark adapted, and even dim objects appear bright. On the other hand, after being in bright sunlight for some time our eyes become bright adapted, and upon going indoors everything looks dark, but gradually we find we can see more and more as our eyes lose their bright adaptation.

The fundamental unit of luminance is the *stilb,* which is defined as one-sixtieth of the luminance of a black body at the temperature of freezing platinum, i.e., about 1770°C. If a luminous surface is *diffusing* (i.e., not shiny), then its luminance is independent of both the area of the surface and the angle of view. Indeed, the constancy of luminance with angle provides the criterion for regarding a surface as being diffusing. If a surface is shiny, or partially specular, we cannot draw any useful conclusions about its luminance, because the luminance of such a surface varies with the direction of view and the presence of light sources above the surface which are partially reflected by it.

The luminance of a surface is a physical property of the surface and does not depend on how it is viewed; indeed, it may not be viewed at all and yet retain its luminance. Viewed from the end of the highway, all the street lamps in a row appear equally bright to our eyes, although the distant lamps appear small and the nearer lamps large. The luminance of a source is quite independent of the size of the source, provided it is uniform over its area. The concept of luminance is not confined to a plane solid surface; the sun, which consists of a ball of glowing gas, has a very high luminance of about 200,000 stilb. A bright spot on the surface of the full moon has a luminance of about 0.25 stilb, with the reflectance (*albedo*) of the moon being about 17%. A sheet of white paper under fairly bright indoor lighting has a luminance of about 0.017 stilb, and under full noon sunlight outdoors may reach about 3.5 stilb. The filament of a tungsten projection lamp may have an effective luminance of 2500 stilb.

It will be seen that the stilb is a fairly bright unit, suitable for measuring light sources. If we are concerned with much dimmer luminances such as some illuminated objects, a more convenient unit is the *nit*, which is defined as one ten-thousandth of a stilb, so that 0.25 stilb is 2500 nit.

B. Intensity

The unit of intensity (I) is the *candela* (formerly *candle power*). The intensity of a source is a measure of its ability to illuminate other objects. If we limit the area of our black body at the temperature of freezing platinum to 1 cm², its intensity will be 60 candelas (cd) in a direction perpendicular to the surface. In any other direction the projected area will be less in proportion to the cosine of the angle from the normal and so will the intensity of the source. This is known as *Lambert's cosine law of intensity*. The intensity of a source also depends directly on its area. Thus,

$$I = BA \cos \theta,$$

where A is the area of the source, B its luminance, and θ the angle of view. If A is 1 cm², B 1 stilb, and θ zero, then the intensity is 1 cd. Thus we may regard the stilb as such a luminance that its intensity is 1 cd/cm² of projected area.

The intensity of an ordinary domestic light bulb varies with direction. The average intensity over a complete sphere is called the *mean spherical candle power,* and it is this which measures the ability of the source to fill a room with light. Many ceiling luminaires for room lighting give light only downward, and the intensity varies considerably with direction. The intensity of a searchlight to anyone standing in the beam could be several million candelas, but this quickly drops to zero as the observer moves out of the beam.

C. Flux

Light *flux* (F) is a measure of the amount of light radiated by a source into a given solid angle. The unit of flux is the *lumen* (lm) defined as the amount of flux radiated by a point source having an intensity of 1 candela into a cone having a solid angle of 1 steradian (sr). The dimensions of flux are those of power (the rate of flow of energy). (Note: The steradian is defined in a manner analogous to the radian in plane angles. The radian is the angle at the center of a circular arc when the length of the arc is equal to the radius of the circle. The steradian is the solid angle at the center of a sphere when the area intercepted by the solid angle on the surface of the sphere is equal to the square of the radius of the sphere. The plane angle of a semicircle is π rad; the solid angle of a hemisphere is 2π sr.)

D. Illuminance

When light falls on an object, it causes that object to be illuminated. This process is called *illumination,* and the quantity of the illumination is called *illuminance* (E). The unit of illuminance is the *phot,* which represents the illuminance produced by 1 lm of light falling on 1 cm^2 of a plane surface. The phot is actually a high level of illuminance, and for ordinary circumstances a more convenient unit is the *lux,* which is the illuminance produced when 1 lm falls on a m^2 of a plane surface. Therefore, there are 10,000 lux in 1 phot.

E. The Inverse Square Law

Suppose a small source having an intensity I cd is at a distance d from a perpendicular screen having a small area A (Fig. 7.1). The solid

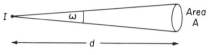

Fig. 7.1. The inverse square law.

angle ω subtended by the screen at the source is A/d^2 sr, hence the flux in the beam is IA/d^2 lm. To determine the illuminance on the screen we must divide the flux by the area of the screen; hence the illuminance is equal to $E = I/d^2$. If I is in candelas and d is in centimeters, the illuminance will be in phots. Similarly, if the distance d is in meters, the illuminance will be in lux. Thus we see that, for a reasonably small source, the illuminance falls off as the square of the distance from the source. This is known as the *inverse square law of illumination*. At one time the illuminance at a distance of 1 m from a small source of 1 candle power was called a *meter-candle*, but the modern term is lumen per square meter (i.e., lux). Similarly, at a distance of 1 cm the illuminance was called a centimeter-candle, but it is now called a lumen per square centimeter, or phot.

There is another Lambert's cosine law for illumination. If a parallel beam containing F lumens falls obliquely on a plane surface, the area on which the light falls is increased by the cosine of the angle of incidence θ. Thus the illuminance is given by

$$E = \frac{F \cos \theta}{A},$$

where A is the cross-sectional area of the incident beam.

F. Photometric Units in the English System

Unfortunately, in the English system of units there are no specific photometric units corresponding to the stilb, lux, etc., in the metric system. Instead, we have to express a given luminance as so many candles per square foot, and an illuminance as so many lumens per square foot (formerly foot-candles). The conversion factors between the English and the metric units depend on the number of square centimeters in one square foot(929) or the number of square feet in one square meter(10.8). Thus 1 stilb is equal to 929 candles/ft² and

G. THE LUMINANCE OF AN AERIAL IMAGE

If we look at a uniform diffusing surface through a window or lens, the image that we see will have a luminance equal to the luminance of the surface multiplied by the transmittance of the window or lens.

In Fig. 7.2, we are looking at a point A of the surface in each case, but at different angles. However, it is a property of a diffusing surface that its luminance is independent of the angle under which it is viewed; hence the luminance at A is the same no matter what kind of optical system may lie between the surface and our eyes. Thus, including a transmittance factor t, we see that

$$B' = tB,$$

where B is the luminance of the diffusing surface and B' the luminance of the image that we see. Although this relation can be proved mathematically, it may well be regarded as a fundamental axiom of photometry.

If we use a lens to form an image of a bright object on the pupil of our eye, the whole aperture of the lens appears to have the same luminance as the source of light, only reduced by the transmittance factor of the lens. This is called the *Maxwellian view*. Similarly, if we look at some object through a telescope, the image that we see will have the same luminance as the object, only reduced by the transmittance factor of the telescope. Under no circumstances can the image seen in a telescope be brighter than the original object. If we look at the moon through a telescope, we get the impression that the image is much brighter than the object, but if we open the other eye and look at the moon directly past the telescope, we see that the real moon is brighter

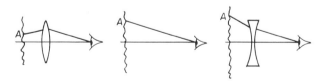

FIG. 7.2. The luminance of a diffusing surface, seen through a window or a lens.

than its image in the telescope. The explanation of this effect is that the telescope accepts more total energy than the unaided eye, which leads to a sensation of dazzle because the image of the moon is so much larger than the original moon.

III. PHOTOMETRIC PROPERTIES OF PLANE SURFACES

A. Fresnel's Formulas and Total Reflection

In 1823 Fresnel developed two remarkable formulas giving the fraction of light reflected at the surface of a sheet of glass at various angles of incidence. If the external angle of incidence in air is I and the internal angle in the glass I', then $\sin I = n \sin I'$, where n is the refractive index of the glass. Fresnel's formulas are as follows:

(a) for light vibrating in a direction parallel to the surface,

$$R_\| = \frac{\sin^2(I - I')}{\sin^2(I + I')}; \tag{1}$$

(b) for light vibrating perpendicular to the surface,

$$R_\perp = \frac{\tan^2(I - I')}{\tan^2(I + I')}. \tag{2}$$

For ordinary unpolarized light we take the average of these two reflectances. In Fig. 7.3, we see two graphs relating the reflectance with the angles of incidence, the angle in air (Fig. 3b) on the right-hand graph and the angle inside the glass (Fig. 7.3a) on the left-hand graph. It will be seen that, for a refractive index of 1.523, for ordinary window glass, the reflectance remains fairly constant up to an angle of incidence (in air) of about 40°, after which it rises at an increasing rate to become 100% at $I = 90°$ at grazing incidence.

Figure 7.3a is important because it shows that the reflection remains fairly constant up to an angle of incidence I' in the glass of about 25°, after which it rises rapidly, reaching 100% at an angle of incidence I' of 41.041°, the sine of I' then being equal to $1/n$. From then on the reflection remains 100% for all angles of incidence; this is known as *total internal reflection*. The angle of incidence in the glass at which the

III. PHOTOMETRIC PROPERTIES OF PLANE SURFACES

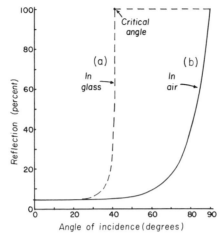

FIG. 7.3. Variation of reflectivity with angle of incidence.

reflection first reaches 100% is the *critical angle*. For angles of incidence less than the critical angle the light is partially reflected, as indicated in Table I. These data are shown plotted in Fig. 7.3.

TABLE I

REFLECTION FROM A GLASS SURFACE[a]

Angle of incidence I' inside the glass (deg)	Angle of refraction I in air (deg)	Reflectivity R (%)
0	0	4.30
5	7.63	4.30
10	15.34	4.31
15	23.22	4.35
20	31.39	4.49
25	40.06	4.89
30	49.60	6.03
35	60.88	9.75
37	66.43	13.67
38	69.66	17.10
39	73.43	22.71
40	78.23	33.60
41	87.68	79.43
41.041	90.00	100.00

[a] For this table, $n = 1.523$

For angles of incidence inside the glass that are greater than the critical angle, the reflectance is indeed 100%, with no losses occurring there. This is a phenomenon used extensively in reflecting prisms (see Section IV Chapter 9) and in fiber optics where thousands of total internal reflections occur without loss of light energy.

The detailed theory of total internal reflection is very complex. The light must leave the denser medium for a short distance to test the refractive index of the lighter medium to see if total reflection should occur, after which it immediately returns into the denser medium. This penetration into the lighter medium is very slight, being only about one wavelength at the critical angle and less than one wavelength at greater angles of incidence in the glass.

B. Brewster's Angle

In 1815, before the announcement of Fresnel's formulas, David Brewster discovered that, at a certain angle of incidence, the light reflected from the surface of a plate of glass is completely polarized. The reason for this is shown in the graphs in Fig. 7.4, which are plotted from Eqs. (1) and (2). At Brewster's angle of incidence B, the whole of the light vibrating perpendicularly to the surface is transmitted and the reflection has dropped to zero, leaving only the light vibrating parallel to the surface in the reflected beam. Thus the reflected light is fully polarized, whereas the transmitted light is partially polarized. Now, Eq. (2) tells us that $R_\perp$ will be zero if $\tan(I + I')$ is infinite, or if $I + I' = 90°$. Hence, at Brewster's angle, the reflected and refracted rays are perpendicular to each other. Also, since $I' = 90° - I$, we know that $\sin I' = \cos I$, and we easily see that $\tan I = n$, the refractive index of the glass. Hence, if $n = 1.523$ for ordinary window glass, the angle of incidence I in air is 56.71°, and the angle of refraction I' in the glass is 33.29°. Brewster windows are frequently used in laser systems, because if the beam is polarized, there will be no transmission loss. The reflectivity for unpolarized light at each surface of a Brewster window is given by Eq. (1), and if $n = 1.523$, the transmission will be 84% per surface, or 72% for the parallel plate.

Because it is sometimes required to rotate the plane of polarization, it is convenient to introduce metallized reflecting surfaces that will not

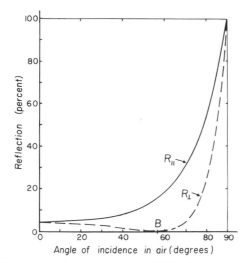

FIG. 7.4. The two polarized components of light reflected from a glass plate.

affect the polarization. One method of doing this is shown in Fig. 7.5. The refractive index of the top prism is chosen such that its Brewster angle is 60°, so that the refractive index must be 1.7320, which is the tangent of 60°. The reflectivity at the polarizing surface is 25%, with the remaining 75% being lost out of the top of the prism. Because the entering and emerging rays are in one straight line, the whole system can be rotated about that line as an axis. The refractive index of the lower prism must be less than 1.50 to ensure total reflection at A. The bottom surface must be metallized because the angle of incidence there is only 30°, which is less than the critical angle.

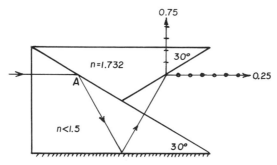

FIG. 7.5. A rotatable polarizing prism.

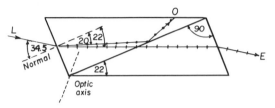

Fig. 7.6. A Nicol prism.

C. The Nicol Polarizing Prism

A much more efficient polarizing prism was suggested by William Nicol in 1828. It is made of calcite, a highly birefringent material, with the refractive indices of the ordinary and extraordinary rays being 1.6585 and 1.4864, respectively, for some wavelength near the middle of the visible spectrum.

The Nicol prism consists of a cleavage rhomb of calcite, cut diagonally by a plane perpendicular to the long diagonal of the end faces, with the two halves cemented together by a layer of Canada balsam, the refractive index of which lies between the two indices of calcite. As indicated in Fig. 7.6, light entering at L is split into two polarized components, with the ordinary ray being reflected to one side, whereas the extraordinary ray passes through the calcite layer and emerges parallel to the incident light. Many variants of the Nicol prism have been devised, some having perpendicular end faces and some a much wider aperture.[1] Today, of course, sheets of Polaroid film are used extensively to form polarized light; they are much cheaper and come in larger sizes.

D. The Wollaston Double-Image Prism

If both of the polarized rays are required, a Wollaston double-image prism is convenient (Fig. 7.7). It consists of two similar quartz prisms cemented together with the optic axes of the two halves being mutually perpendicular. Thus the ordinary ray in the first prism becomes the extraordinary ray in the second, and vice versa. If the slope of the

---

[1] J. M. and H. E. Bennett, Polarization, in "Handbook of Optics" (W. G. Driscoll and W. Vaughan, eds.), Section 10. McGraw Hill, New York, 1978.

III. PHOTOMETRIC PROPERTIES OF PLANE SURFACES

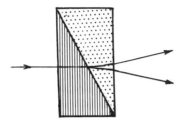

FIG. 7.7. The Wollaston double-image prism.

interface is correctly chosen, the two emerging rays will be about equally inclined to the axis.

E. Reflection and Transmission at Normal Incidence

If the light is incident perpendicularly to the surface, both of Fresnel's formulas degenerate to

$$R = \left(\frac{n-1}{n+1}\right)^2.$$

The magnitude of this reflection increases rapidly for high-index materials (as seen in the following tabulation).

n	1.5	1.6	1.7	2.0	3.0	4.0
R	0.040	0.053	0.067	0.111	0.250	0.360

This tabulation gives the reflectance at one surface. For a glass plate, the reflection occurs at each surface, but the transmission losses are not quite equal, for two reasons—first, less light reaches the second surface because some of it has already been lost at the first surface, and second, interreflections within the plate lead to a slight increase in the amount of light transmitted.

The actual transmittance of a plate of glass can be calculated by summing an infinite geometric series. The sum of any geometric series is given by

$$S = A/(1 - r),$$

where A is the first term and r the common ratio.

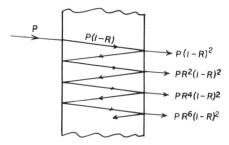

Fig. 7.8. Interreflections within a glass plate.

The optics case is illustrated in Fig. 7.8, which shows a parallel plate of nonabsorbing material with a surface reflectivity R and consequently a surface transmittance $1 - R$. The incident light is P and the light emitted by each of the multiple transmitted beams is, in order, $P(1 - R)^2$, $PR^2(1 - R)^2$, $PR^4(1 - R)^2$, and so on. The sum of this geometric series is

$$S = \frac{P(1 - R)^2}{1 - R^2} = P\left(\frac{1 - R}{1 + R}\right).$$

The result of using this formula is shown in the following tabulation:

Refractive index n	1.5	1.6	1.7	2.0	3.0	4.0
Transmission ignoring interreflections	0.922	0.897	0.870	0.790	0.562	0.410
Transmission including interreflections	0.923	0.899	0.874	0.800	0.600	0.471

Since surface reflection is a pure function of refractive index, the transmission of a parallel plate must also be a function of refractive index. Substituting $R = (n - 1)^2/(n + 1)^2$ in the expression for the transmission S, we find that, including interreflections,

$$S = P\left(\frac{2n}{n^2 + 1}\right).$$

Actually, this problem can be handled in an entirely different and much more general way. Because of the interreflections there will be two streams of light inside the plate, one going from left to right in the direction of the incident light and a much smaller stream going in the

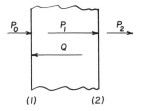

FIG. 7.9. Two streams of light within a glass plate.

opposite direction, as indicated in Fig. 7.9. Remembering that at a surface the reflected fraction is R and the transmitted fraction is $1 - R$, we see that

$$P_1 = P_0(1 - R) + QR \quad \text{at surface (1);}$$
$$P_2 = P_1(1 - R) \quad \text{at surface (2);}$$
$$Q = P_1 R.$$

Eliminating P_1 and Q from these three equations gives the plate transmittance as

$$P_2 = P_0\left(\frac{1 - R}{1 + R}\right),$$

as before. The reflectance of the plate is one minus the transmittance, or $2R/(1 + R)$. This general procedure can be extended to a pile of plates, all having the same refractive index, giving the final transmittance as

$$\frac{1 - R}{1 + (N - 1)R},$$

where N is the number of surfaces in the pile. The reflectance of the pile is one minus the transmittance, or

$$\frac{NR}{1 + (N - 1)R}.$$

F. Antireflection Coatings

Dating from approximately 1938, almost all lenses have been equipped with an antireflection coating on the various lens surfaces. This is an evaporated thin film of transparent dielectric material which, if properly applied, has the effect of reducing the reflectivity of the surface, and hence improving the transmittance because no light is lost at a transparent layer. Reduction of surface reflection has two advantages—it improves the light transmission and reduces the brightness of ghost images caused by interreflections between the lens surfaces.

The action of an antireflection coating is indicated in Fig. 7.10. Incident light(ray *b*) is reflected at the upper surface of the coating and also at the interface between the coating and the substrate(ray *a*). If the refractive index of the coating material is equal to the square root of the refractive index of the substrate, the intensity of the two reflected beams will be equal. Furthermore, if the optical thickness of the coating is equal to one-quarter of the wavelength of the light, the phase of the upper reflection will be opposite to the phase of the interface reflection, and the two reflected beams will interfere and produce darkness at *R*.

It is harder to understand why the transmission is increased by an antireflection coating. There will be two beams in the transmitted light, ray *b* passing directly through the coating and ray *a*, reflected twice, once at the coating–substrate interface and again at the upper surface of the coating where it suffers a 180° phase reversal. Consequently, when these two beams meet at *T*, they will be in phase and their intensities will add, whereas in the reflected beam at *R* the two beams are in opposite phase and so lead to destructive interference.

Because the effect of the coating depends on the wavelength of the

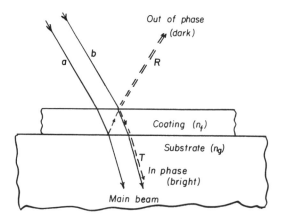

Fig. 7.10. Action of an antireflection coating.

light, it is perfectly effective at only one wavelength; however, it is almost as effective at other wavelengths throughout the visible spectrum. For a greater degree of reflection reduction, a multilayer coating is often used, which is more effective throughout the spectrum and also more effective if the square root of the substrate index lies outside the range of refractive indices of available coating materials.

G. Beam Splitters

In many instrumental applications it is desired to divide a beam of light into two separate beams emerging in different directions. In parallel light, the simplest form of beam splitter is a plate of glass set at 45° in the beam, with one surface of the plate made reflective by depositing on it a thin transparent layer of metal, or better, a multilayer dielectric coating which can either be achromatic or can divide the spectrum between the two beams, sending one color one way and its complementary color the other way.

Trouble may arise if a thick beam splitter is used in converging light. The reflected beam may show a double image and the transmitted beam exhibit astigmatism. It has been proposed to form a weak convex surface on the rear of a beam splitter to correct the aberrations of the transmitted beam, or alternatively, to follow the beam splitter with a meniscus lens element.[2] These effects can be eliminated by making the beam splitter very thin (a stretched pellicle reflector is often used). A better arrangement is to deposit the reflective material on the hypotenuse of a 45° prism and then cement an identical prism to it, thus forming a beam-splitting cube. It is necessary to include the effect of the thick glass plate in the design of the lens using the cube.

In cameras for color television it is necessary to divide a beam three ways, with dichroic reflectors to send the red, green, and blue components of the image into three separate Vidicon tubes. A possible arrangement is shown in Fig. 7.11. A dichroic film on surface *a* reflects blue light and transmits the rest. Another film at *b* reflects red and transmits green. Care must be taken to see that the glass path in each of the three beams is the same length, so that its aberrational effect can be designed into the relay or zoom lens used in the camera.

---

[2] V. J. Doherty and D. Shafer, Simple method of correcting the aberration of a beamsplitter in converging light. *Proc. SPIE* **237**, 195 (1980).

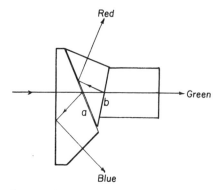

FIG. 7.11. A three-way color television beam splitter.

H. Density

The density of a transparent layer such as a photographic negative is defined as the log of the ratio of the incident flux (F_{inc}) to transmitted flux (F_{trans}), or

$$\text{density} = \log_{10}\left(\frac{F_{inc}}{F_{trans}}\right).$$

There is, oddly, no name for the unit of density. A film having a density of 1.0 transmits 10% of the incident light, whereas a film with a density of 2.0 transmits 1.0% of the light. Densities as high as 5.0 have been measured; in that case the film transmits only 1/100,000 of the incident radiation (see Table II).

A photographic film is transparent but, because of its grainy structure, it tends to scatter some of the incident light into a wide cone. Thus the numerical value of the density of a particular piece of film will depend on the amount of scattered light that is included in the "transmission." A typical graph of density versus angle of pickup is shown in Fig. 7.12. If none of the scattered light is included, the density is said to be *specular*, whereas if all the scattered light is included the density is *diffuse*. By the nature of density, it is clear that the specular density has the maximum value whereas the diffuse density is a minimum for a given film.

If a photographic film is projected by a specular enlarger or slide projector, it is the specular density that is significant. On the other

III. PHOTOMETRIC PROPERTIES OF PLANE SURFACES

TABLE II
Relation Between Transmission and Density

Transmission (%)	Density
100	0
90	0.046
80	0.097
70	0.155
60	0.222
50	0.301
40	0.398
30	0.523
20	0.699
10	1.000
9	1.046
.	.
.	.
.	.

hand, if the film is contact printed or enlarged with a sheet of opal glass behind the film, the diffuse density is important. In the absence of other comment it is generally the diffuse density that is meant. Truly specular density is rarely encountered because some of the scattered light is almost always picked up by the receiver.

I. Densitometers

A *densitometer* is an instrument for measuring density, and generally it is the diffuse density that is measured, as there is an illuminated

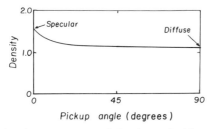

FIG. 7.12. Relation between measured density and pickup angle for a typical photographic negative emulsion.

opal glass behind the film and a small photocell in front of it. Only a small area of the film is involved in this measurement.

If it is desired to plot the variation of density from point to point along a film image, for instance along a photograph of a spectrum, then a microdensitometer is used.[3] In this instrument, a narrow slit illuminated from behind is imaged on the film by a microscope objective, with the dimensions of the system such that a line of light about 1/1000 in.(25 μm) wide is formed on the film. Below the film is another microscope objective that projects the line of light into a second slit wider than the first, serving as a stray-light shield; behind this second slit is a photocell that measures the light passing through the film. The film is moved slowly along past the line of light, and a recording device plots a graph of either the density or the transmittance of the image (whichever is required). The density measured in this way is nearly equal to the diffuse density because the microscope objective emits a finite cone of light, the angle of which depends on the numerical aperture of the objective (as seen in the following tabulation).

Numerical aperture	0.25	0.40	0.65	0.85
Cone angle (deg)	±14	24	40	58

IV. PHOTOMETRIC MEASURING INSTRUMENTS

A. The Eye as a Photometer

The eye is a sensitive but very limited photometer; the only photometric determination the eye can make is whether two adjacent white areas are or are not equally bright. Even this determination will be precise only if the dividing line between the luminous areas is vanishingly narrow and if the luminances are at a medium level (not too bright and not too dim), especially if the two areas have the same color. This is all that the eye (as a photometer) can do. We cannot say that

[3] R. E. Swing, Microdensitometer optical performance. *Opt. Eng.* **15**, 559 (1976).

one area is, for example, five times as bright as another, or even whether two areas, widely separated and with a different luminosity between them, are equally bright. Some observers claim to be able to match the luminosity components of adjacent areas having different colors, but this falls into the realm of colorimetry and will not be discussed further here.

As photometry is basically related to the eye, the fundamental photometric instruments must be visual. However, as visual photometry is slow and tedious, many instruments have been developed in which a photocell is used to replace the eye as the detector. Of course, an appropriate light filter is needed in front of the cell to ensure that the spectral response of the cell matches that of the eye. If the instrument using a photocell is properly constructed, it will give the same result as the slower and more cumbersome visual instrument, and, moreover, a photoelectric instrument is able to determine that one luminance is five times as great as another luminance widely separated from it, something that the eye cannot do.

B. Intensity Photometers

An intensity photometer is an instrument for measuring the candle power of a light source. This requires the existence of a source of known candle power for comparison with the unknown source. The two sources are mounted on a long optical bench with two small mirrors and a diffusing screen between them, as indicated in Fig. 7.13. One or both of the sources is moved along the bench until the luminances of the two sides of the diffusing screen appear equal to the observer. The eye is remarkably sensitive to the matching of luminances — provided the line of separation between the areas is narrow and the sources have the same color. Under these conditions, equal luminances imply equal illuminances, and since illuminance is given

Fig. 7.13. An intensity(candle power) photometer.

by the inverse square law, we see that at balance

$$I_1/d_1^2 = I_2/d_2^2,$$

so that if one intensity is known, the other can be immediately found.

The original intensity standard is a black body at the temperature of freezing platinum, and this is maintained at the national standards laboratories. Secondary standards are supplied by these laboratories for inductrial use in everyday photometry.

Because most light sources are nonuniform in intensity, it is usual to measure the candle power in several different directions and with the source tilted at several different angles. The "mean spherical candle power" of a source can be measured with the help of an integrating sphere (see Part E).

C. Illuminance or Flux Meters

Today, most illumination meters are photoelectric, i.e., incident light flux falling on a photocell produces a measurable electric current or voltage. The dial of such an instrument must be calibrated by exposing the cell to a series of known illuminances and marking the dial accordingly. The markings may refer to the flux falling on the cell in lumens or to the illuminance on the cell in phot, lux, or foot-candles (since it is known that illuminance is equal to flux divided by area). Such a cell can sometimes be used to measure the flux in a beam that does not fill the area of the cell; in other cases the reading will be accurate only when the whole area of the cell is uniformly illuminated. Illumination meters of this kind are often used to measure the illumination on a desk top or factory bench.

D. Luminance Meters (Telephotometers)

A luminance meter may be visual or photoelectric. In a visual instrument, the luminance of an image of the source to be measured is compared with a known luminance inside the instrument. In a photoelectric instrument, the flux from an image of the source passing through a small hole is measured with a photocell.

IV. PHOTOMETRIC MEASURING INSTRUMENTS   113

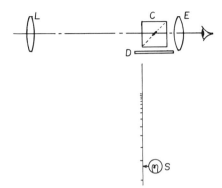

FIG. 7.14. A visual telephotometer.

(a) *Visual telephotometers.* In a visual luminance-measuring instrument, an image of the object to be measured is formed by a lens L in a beam-splitting cube C (Fig. 7.14). At the middle of the cemented interface in the cube is a small reflecting spot at 45°, located at the focus of the lens L and also at the focus of the eyepiece E, the lenses L and E thus constituting a small telescope. The reflecting spot reflects light into the eye from a diffusing screen D which is illuminated from below by a sliding lamp S. In use, the instrument is directed so that the image of the object to be measured falls alongside the reflecting spot, and the lamp S is raised or lowered until the luminance of the spot matches that of the object as seen in the eyepiece. The luminance scale of such an instrument cannot be calibrated by calculation because there are too many unknown quantities involved. It must, therefore, be calibrated experimentally by observing two or three known luminances and by interpolation of intermediate markings using the inverse square law. Such an instrument can be used as an illuminance meter by observing the luminance of a diffusing tablet having a known reflectivity, and in some cases the scale of the sliding lamp is calibrated directly in illuminance units.

(b) *Photoelectric telephotometers.* Today, most luminance photometers are photoelectric. The basic arrangement of such an instrument is indicated in Fig. 7.15. The lens L forms an image of the test object on the surface of a metal screen having a small hole on its axis. Behind the hole is mounted a photocell (PEC) to measure the flux passing through the hole. The dial of the meter is then calibrated to

7. THE PHOTOMETRY OF OPTICAL SYSTEMS

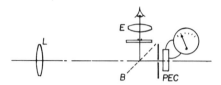

FIG. 7.15. A photoelectric telephotometer.

read luminance directly. In use, some kind of viewfinder is required to ensure that the instrument is measuring the desired point on the object. This may consist of a beam splitter B which reflects an image of the scene into the focal plane of an eyepiece E, with a little circle on a reticle in the focal plane indicating the exact location of the hole in front of the photocell. In some instruments the dial of the electric meter appears in the eyepiece field so that the observer can read off the luminance of any part of the scene without removing the instrument from his eye.

It should be remarked that both these instruments assume the truth of the relation $B' = tB$ and that the luminance of an image is proportional to the actual luminance of the object.

E. The Integrating Sphere

The integrating sphere is a familiar piece of laboratory equipment. It consists of a large hollow sphere lined with highly reflective diffusing white paint, with the property that any flux admitted into the sphere becomes uniformly distributed over the entire sphere wall and the luminance so produced is strictly proportional to the admitted flux. The flux may come from a lamp whose lumen output is to be measured or from an admitted beam of light from an outside source such as a slide projector. The luminance of the sphere wall can be measured by means of a telephotometer looking through a small hole in the wall or by a photocell set in the sphere wall in such a position that no direct flux falls on it. In order that the proportionality is accurate, it is essential that any holes in the sphere wall are very small (less than 2% of the sphere area) and that there are no obstructions inside the sphere. The effect of lamp holders, etc., can be minimized by painting them with the same paint used on the sphere wall.

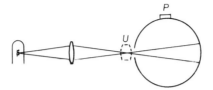

FIG. 7.16. Arrangement for measuring the transmittance of a lens.

The exact relation between admitted flux and wall luminance is unknown, and only comparative measurements are possible. A lamp of known lumen output can be compared with an unknown lamp, or the flux in a light beam can be compared with a known flux.

F. Transmittance Measurement

The transmittance of a plate of glass or large lens can be measured directly by viewing a luminous diffusing surface through a telephotometer, either with or without the plate in between. This follows from the law $B' = tB$.

A small lens is harder to measure and may require the use of an integrating sphere, as indicated in Fig. 7.16. A small lamp filament is imaged into the test lens U, and the transmitted light enters the integrating sphere. The photocell P is read twice, with and without the lens in place, and the ratio found. It is essential that the beam of light passing through the test lens be so narrow that no light is obstructed by the lens mounts or other apertures.

The transmittance of a telescope[4] can be measured directly by placing the exit pupil close to a uniform diffusing source and then comparing the luminance of the source itself with its image in the aperture of the telescope objective.

The measurement of the transmittance of a microscope is a much more difficult operation. It requires the construction of an "electric eye" having a very small entrance aperture, say, less than 1 mm in diameter, backed up by a photocell subtending a limited field angle, say, $\pm 10°$, at the entrance pupil (Fig. 7.17a). A small photocell could

---

[4] J. Guild, Apparatus for the photometry of optical instruments, *in* "Optical Instruments; Proceedings of the London Conference 1950" (W. D. Wright, ed.), p. 201. Chapman and Hall, London, 1951.

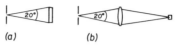

FIG. 7.17. An "electric eye" for measuring the transmittance of a microscope.

be used by imaging the entrance pupil onto the photocell by means of a field lens, as shown in Fig. 7.17b. To measure the transmittance of a microscope, this "eye" is mounted above the eyepiece of the microscope, taking care that the pupil of the "eye" fits exactly into the exit pupil of the microscope, both longitudinally and laterally. The microscope is focused onto a white diffusing surface and the photocell reading is made with the "eye" in place, and then with the "eye" resting on the diffusing source after removing the microscope. In this way, the same aperture and field are employed in the measurement, both with and without the microscope.

G. Spectrophotometers

A spectrophotometer is an instrument for plotting the wavelength distribution of light transmitted or reflected by a colored object. Light from a convenient source is passed through a monochromator to provide a beam containing only a narrow band of wavelengths. This beam is then split into two, one-half being transmitted by the specimen while the other half passes through some means for varying the intensity in a measurable manner.

In a visual spectrophotometer, the two beams are brought into a pair of adjacent visual fields so the observer can achieve a luminance match by varying the intensity of the second beam. He then changes the wavelength and repeats the measurement. It is customary to make two runs, one with the specimen in the beam and one without, and then to plot the difference between readings versus wavelength. The reflectance of a colored object can, of course, be plotted as easily as the transmittance of a transparent object. The whole optical system can be operated in reverse order, if preferred.

In an automatic recording spectrophotometer, the two beams are presented alternately to a photocell by means of a rotating chopper, the output from the cell thus containing a dc and an ac component. The dc component is rejected, and the ac component is phased against the rotating chopper so that it can be used to drive both the means for varying the beam intensity and the recording pen in the y direction. The graph paper is moved in the x direction by the wavelength drum of the monochromator.

H. Photographic Exposure Meters

A simple hand-held exposure meter such as the well-known Weston instrument measures the average luminance of the scene, limited, it is hoped, to the angular field that will be accepted by the camera. A slide rule is supplied with the instrument by means of which the exposure time can be calculated for any given film speed and stop aperture. It may be necessary to make various corrections (such as the change in the effective F-number with a near object) and reasonable assumptions as to the lens transmittance. A meter mounted on top of the camera serves the same purpose.

A through-the-lens meter is an entirely different matter. This meter does not measure the object luminance; instead it measures the illumination that will fall on the film, thus determining the correct exposure time without needing any adjustments. In many SLR cameras the film speed dial is mounted on top of the shutter speed control, and a photocell mounted above the ground-glass screen measures the illumination there. By matching pointers, or by some other device, the correct exposure is immediately determined. In automatic cameras the stop aperture is adjusted electrically to give the right exposure for any pre-set shutter speed.

V. THE FLUX EMITTED BY A PLANE SOURCE

A. Flux Radiated into a Cone

We have seen that the flux radiated into a cone from a point source is $F = I\omega$, where I is the intensity of the source in candelas and ω is the solid angle of the cone in steradians (the flux then being in lumens). However, the intensity of a plane source falls off at increasing angles

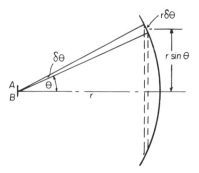

FIG. 7.18. Flux radiated into a cone from a small plane source.

because of Lambert's cosine law of intensity, so that the flux in a cone is less than it would be if the source were radiating equally in all directions. To determine the actual flux from a plane source requires integration.

Figure 7.18 shows a small plane source of area A and luminance B, radiating into a cone of semivertical angle θ. A sphere with radius r is drawn about the source as center, and we imagine a pair of coaxial cones having semiangles θ and $\theta + \delta\theta$, respectively, with their axes perpendicular to the plane of the source.

We now proceed to determine the flux radiated into the conical shell between the two cones and then to integrate this over the whole cone. The intensity in the direction of the shell is $I_\theta = AB \cos \theta$ candelas, and the solid angle of the shell is equal to the area intercepted on the surface of the sphere divided by the square of the sphere radius, or

$$\omega = (2\pi r \sin \theta) r \, \delta\theta / r^2 \quad \text{sr}.$$

Hence, the flux radiated into the shell is the product of intensity and solid angle, or

$$\delta F = AB \cos \theta \left(\frac{2\pi r^2 \sin \theta \, \delta\theta}{r^2} \right) = \pi AB (2 \sin \theta \cos \theta \, \delta\theta) \quad \text{lm}.$$

Integrating this over a finite cone of semiangle θ gives

$$\text{flux in cone} = \pi AB \int_0^\theta 2 \sin \theta \cos \theta \, d\theta = \pi AB \sin^2 \theta \quad \text{lm}. \quad (3)$$

This is an important result having many applications.

B. The Lambert

By writing $\theta = 90°$, we see that the flux radiated by a small plane source into a complete hemisphere is given by

$$F = \pi AB \quad \text{lm}.$$

The term used to express the amount of flux radiated into a hemisphere is the *lambert* (L) and it is defined as the flux radiated into a hemisphere by 1 cm² of a diffusing surface having a luminance of 1 stilb. Thus, if the surface area is A cm² and its luminance is B stilb, the hemisphere radiation is given by

$$F = \pi AB = AB_L \quad \text{lm},$$

where B_L is the surface luminance in lamberts, and $B_L = \pi B$.

If the unit of length is the meter, the unit of luminance is the nit and the total radiation is in meter-lamberts (mL) or apostilbs. In the English system, the unit of luminance is the candle per square foot and the total hemisphere radiation is in foot-lamberts (fL).

Over the years the true significance of the lambert has been lost, and it is now customary to regard the lambert as just another unit of luminance, where the number of lamberts is π times the ordinary luminance in candles per unit area. Thus a surface having a luminance of, say, 200 candles/ft² is said to have a luminance of $200\pi = 628$ fL.

The lambert concept is strictly applicable to a diffusing surface, but it is common to have the luminance of a partially specular surface such as a projection screen expressed in foot-lamberts.

It is useful to remember that there are 929 cm² in 1 ft², and hence the foot-lambert is only slightly larger than the millilambert. The latter term is often used in scientific literature as a unit of luminance, although the foot-lambert is more common in this country.

C. The Luminance of an Illuminated Surface

We must now determine the relation between the illuminance falling on a surface and the luminance resulting from this illumination.

Suppose a plane receiving surface of area A is under an illuminance of E lumens per square unit. The total number of lumens falling on the surface is therefore equal to AE. If the surface is perfectly diffusing and has a reflectivity r, then a fraction r of the incident light is scattered into a hemisphere. Thus, if the resultant lambert measure of the surface is B_L, we have

$$rAE = AB_L,$$

or

$$B_L = rE.$$

The luminance of the surface in lamberts is therefore equal to the product of the reflectivity and the illuminance falling on the surface. This is the chief reason for the popularity of the lambert as a measure of luminance. Because the reflectivity of ordinary white paper is nearly 1.0, the luminance of a sheet of white paper in lamberts is numerically equal to the illumination falling upon it, and white paper under an illumination of 1 fc has a luminance of approximately 1 fL. The reflectivity of most ordinary objects is less than 1.0, but some specular projection screens have a "gain" of $r = 2.0$ or 3.0, or even more, and in such cases the luminance in lamberts is greater than the illuminance falling on the screen, at least when viewed along the normal to the screen surface.

VI. ILLUMINANCE DUE TO A CIRCULAR SOURCE

A. Axial

An important relation is photometry is an expression for the illuminance due to a large circular source at a point on its axis. We shall assume that the source is a perfect diffuser with luminance B and that its radius subtends an angle θ at the center of the screen (Fig. 7.19).

This problem can be handled in two ways. The direct approach is to divide the source into a number of annular zones of radius r and width dr, each zone subtending an angle $d\theta$ at the axial point on the screen. The intensity of such a zone in the direction of E_0 is $2\pi r\, dr\, B \cos \theta$ candelas, with the factor $\cos \theta$ coming from Lambert's cosine law of intensity. Since $r = x \tan \theta$, we see that $dr = x \sec^2 \theta\, d\theta$, and the

VI. ILLUMINANCE DUE TO A CIRCULAR SOURCE

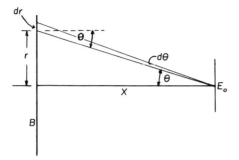

FIG. 7.19. Illumination from a large circular source at a point on the axis.

illuminance at E_0, by the inverse square law, is

$$dE = \frac{2\pi B(x \tan \theta)(x \sec^2 \theta \, d\theta)\cos^2 \theta}{x^2 \sec^2 \theta}.$$

The second $\cos \theta$ in the numerator comes from Lambert's cosine law of illumination. This expression for the zonal illuminance reduces to

$$dE = 2\pi B \sin \theta \cos \theta \, d\theta,$$

and when integrated from 0 to θ gives

$$E_0 = \pi B \sin^2 \theta. \tag{4}$$

The illumination at the center of the screen is therefore independent of the distance of the source from the screen and depends only on the luminance of the source and the angle it subtends at the screen.

The second way to evaluate E_0 is to imagine a small hole of area A existing at the axial point of the screen. The flux passing through this hole will be given by Eq. (3) to be

$$F = \pi AB \sin^2 \theta.$$

Obviously, the flux passing through the hole will be equal to the flux falling on it, and if we imagine the hole to be filled with screen material, we see that the illuminance at that point on the screen will be equal to the flux divided by the area, or

$$E_0 = \pi B \sin^2 \theta,$$

as before. If the source is of infinite angular extent, such as the open sky or the interior of an integrating sphere, the angle θ becomes 90° and the illuminance becomes

$$E_0 = \pi B = B_L,$$

so that the illuminance is then numerically equal to the luminance of the source in lamberts, independent of the distance of the screen from the source. Thus there is no inverse square law if the source is of infinite extent.

B. Oblique

(*a*) *A small source; the cos⁴ law.* If a plane diffusing self-luminous source of light is small compared with its distance from a screen placed parallel to the source, it is easy to see that the illumination on the screen will be a maximum on the axis of the source and will diminish at increasing obliquity angles ϕ.

In Fig. 7.20, we see that if X is the axial distance from source to screen, the oblique distance at angle ϕ is $X \sec \phi$, and by the inverse square law, the illumination at angle ϕ will be inversely proportional to $(X \sec \phi)^2$. In addition to this loss, there are two Lambert's cosine laws that have to be taken into account, one at the source and the other at the screen. The net result is that

$$E_\phi = E_0 \cos^4 \phi.$$

This is the well-known cos⁴ law of illumination.

Since we can regard the exit pupil of a lens as being a diffusing disk, at least to first approximation, we see that the cos⁴ law can be applied

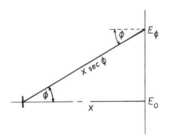

FIG. 7.20. The cos⁴ law.

roughly to the distribution of illumination over the field of a photographic lens.

(b) *A large circular source.* If the self-luminous diffusing disk considered in the last paragraph is large compared with its distance from the screen, the obliquity angle ϕ differs from point to point over the source, and it is necessary to employ integration to determine the true illuminance at any point on the screen. The problem is a difficult one, but it has been solved by Foote.[5] His formula for the oblique illuminance at a point on the screen having an obliquity angle ϕ from the midpoint of the source, the size of the disk being specified by the subtense U' of the disk radius at the middle of the screen, is

$$E_\phi = \frac{\pi B}{2}\left\{1 - \frac{\sec^2 \phi - \tan^2 U'}{[\tan^4 \phi + 2\tan^2 \phi(1 - \tan^2 U') + \sec^4 U']^{\frac{1}{2}}}\right\}.$$

This formula degenerates to

$$E_0 = \pi B \sin^2 U'$$

when ϕ is zero. This is the familiar relation given in Eq. (4).

Actually, Foote's formula does not differ greatly from the simple $\cos^4$ law. A few typical values of the ratio E_ϕ/E_0 are given in Table III. For small values of U', for example, when $U' = 3.58°$(which is the marginal ray slope for an $f/8$ lens), the true and $\cos^4$ values are identical, at least to $\phi = 40°$ from the axis. For larger disks, the illumination increases slightly over the $\cos^4$ value.

TABLE III

VALUES OF E_ϕ/E_0

ϕ (deg)	$\cos^4\phi$	$U' = 7.2°(f/4)$	$U' = 10.3°(f/2.8)$	$U' = 14.5°(f/2)$	$U' = 20.9°(f/1.4)$
			E_ϕ/E_0		
0	1	1	1	1	1
10	0.941	0.942	0.944	0.947	0.954
20	0.780	0.785	0.791	0.801	0.822
30	0.562	0.570	0.577	0.592	0.625
40	0.344	0.351	0.357	0.370	0.401

[5] P. D. Foote, Illumination from a radiating disk. *Bull. Bur. Std.* **12**, 583 (1915).

VII. ILLUMINATION IN AN OPTICAL IMAGE

A. Axial, with Distant Object

Suppose we have a lens forming an image of a distant object on a screen, as in an ordinary camera. If the object is diffusing, with a uniform luminance of B candles per unit area over its entire surface, then the exit pupil of the lens will be filled with light having this same luminance B (except for transmission losses), and the illuminance at the axial point of the screen will be given by Eq. (4) to be

$$E_0 = \pi B \sin^2 \theta,$$

(see Fig. 7.21). Now it can be shown that if the object is very distant and the lens is perfectly corrected for spherical aberration and coma, as all good lenses are, the second principal plane is actually part of a sphere centered about the focal point F_2. Hence

$$\sin \theta = Y/f' = D/2f',$$

where D is the diameter of the entering beam. Thus, for a perfectly corrected lens with a distant object,

$$E_0 = \pi B D^2 t / 4 f'^2,$$

where t is the transmittance factor of the lens.

For nearly 100 years it has been customary to express the ratio of the focal length of a lens to the diameter of the entering beam as the F-number of the lens (symbol N), where

$$N = 1/(2 \sin \theta) \quad \text{or} \quad \sin \theta = 1/2N.$$

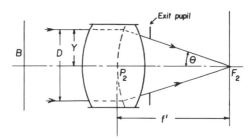

Fig. 7.21. Illuminance in the image of a distant object.

TABLE IV

SOME WELL-KNOWN FRACTIONS OF A STOP

Fraction	Area ratio	Diameter ratio
Whole	2 (or 0.5)	1.414 (or 0.707)
Half	$\sqrt{2} = 1.414$	1.189 (or 0.841)
Third	$\sqrt[3]{2} = 1.260$	1.122 (or 0.891)
Quarter	$\sqrt[4]{2} = 1.189$	1.091 (or 0.917)
Sixth	$\sqrt[6]{2} = 1.122$	1.059 (or 0.944)

Hence, the expression for the image illuminance becomes

$$E_0 = \pi B t / 4 N^2.$$

Since photographic exposure is proportional to the product of illumination and time, it will be inversely proportional to the square of N. Thus, halving the F-number, say, from 5.6 to 2.8, will require one-quarter the exposure time to produce the same density on the film.

It is common among photographers to refer to certain "fractions of a stop." A whole stop is an aperture change to one which has double or half the aperture area, or a factor of $\sqrt{2}$ or $1/\sqrt{2}$ times the aperture diameter. Other well-known fractions of a stop are shown in Table IV.

B. Axial, with Near Object

Figure 7.22 illustrates the situation existing when the object is close to the lens. The image is now at a distance from F_2 given by mf', where m is the image magnification. Similarly, the distance of the exit pupil

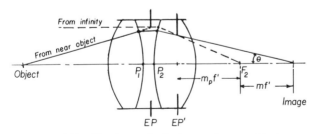

FIG. 7.22. Illuminance in the image of a near object.

from F_2 is given by $f'm_p$, where m_p is the ratio of the diameter of the exit pupil to the diameter of the entrance pupil, i.e., the pupil magnification. Thus, if r and r' are the radii of the pupils,

$$\sin \theta = \frac{r'}{f'(m + m_p)} = \frac{rm_p}{f'(m + m_p)}$$

$$= \frac{D}{2f'(1 + m/m_p)} = \frac{1}{2N(1 + m/m_p)}$$

The effective F-number is now equal to the true F-number multiplied by $1 + m/m_p$, and the image illumination on axis is given by

$$E_0 = \frac{\pi Bt}{4N^2(1 + m/m_p)^2}.$$

If the lens is symmetrical, or nearly so, about a central stop, the pupil magnification will be approximately 1.0 and it can be ignored, but in a telephoto or reversed telephoto the pupil magnification may range from perhaps 0.5 to perhaps 2.0, and it must be taken into account. However, such lenses should never be used with a near object, so the problem does not often arise. Of course, in an SLR camera with through-the-lens metering, these factors are taken care of automatically.

C. The Exposure Equation

During World War II, the American Standards Association [now the American National Standards Institute (ANSI)] established a measure of film speed that has become international in scope. It is based on the relation that if the time of exposure required to produce a satisfactory picture is T seconds, then

$$T = N^2/BSt,$$

where N is the F-number of the lens aperture, B the object luminance in candelas per square foot, S the ASA film speed, and t the lens transmittance. Thus, for example, if $N = 8$ (for an $f/8$ lens), $B = 200$ (for an average outdoor landscape), $S = 64$, and $t = 0.9$, the required exposure time is

$$T = 0.00556 = 1/180 \quad \text{sec}.$$

A well-known rule of thumb states that for an average outdoor scene in bright sunlight, the exposure time will be about equal to the reciprocal of the ASA film speed at $f/16$. This fits directly into the exposure equation, because a scene luminance of 200 is about equal to the square of 16 ($=256$).

In Germany and some other European countries, film speeds are expressed in the DIN system. The relation between speeds in the DIN and ASA systems is logarithmic, being

$$S_{DIN} = 10 \log S_{ASA} + 1,$$

so that an ASA speed of 400 becomes 27 in the DIN system, and ASA 64 becomes DIN 19, and so on.

At one time, many camera lenses were marked in the so-called Uniform Scale (U.S.) system. The virtue of this system was that the exposure would have to be directly proportional to the U.S. number rather than to its square, with the conversion factor between the F system and the U.S. system being

$$F\text{-number} = 4\sqrt{\text{U.S. number}}$$

or

$$\text{U.S. number} = (F\text{-number})^2/16.$$

Thus $f/4$ becomes U.S. 1, and $f/16$ is U.S. 16.

D. The T-Stop System

In professional motion-picture photography, it is necessary to ensure that a scene taken through a variety of different lenses will always have the same density, at least in the middle of the field. For this reason, such lenses are often engraved with photometrically-determined aperture markings in which the lens transmittance is embodied in the stop number by use of the so-called T-stop, where

$$T\text{-stop} = (F\text{-number})/\sqrt{t},$$

so that the exposure equation becomes

$$\text{Time } T = (T\text{-number})^2/BS.$$

The need for T-stops has largely disappeared as a result of the universal use of antireflection coatings and color-free glass; they are still needed in special cases such as when a beam splitter is included in the optical system, the effective transmittance then being perhaps only 0.5 or less.

E. Stray Light in Lenses

Stray (unwanted) light in an optical image comes from two independent sources. It may arise from mechanical scattering on the inside of the lens barrel or from the ground rims of the lens elements; it may also arise from interreflections between pairs of lens surfaces. In the latter case, since the surfaces are spherical and highly polished, a series of images of a bright source such as the sun is formed, even if the source itself is outside the picture area. These so-called ghost images are often seen in motion pictures or television; they all lie on a line joining the bright source to the middle of the picture area. Most of these ghost images will be out of focus, and in any case they are afflicted with every possible aberration and appear as irregular blobs of light.

Occasionally, when a lens is stopped down to a very small aperture, an image of the iris diaphragm is formed by reflection in the middle of the picture. This image may be in focus, or it may be so out of focus that only a vague patch of light is formed. This is generally called a *flare spot*.

If the stray light does not form any recognizable ghost images or flare spots, it is generally referred to as *veiling glare*. This has the effect of lowering the overall contrast of the image. It can be measured by forming an image of the inside of a light box with a black absorbent disk in the middle of the opposite wall. A comparison of the illuminance in the image of the disk with that of the light background provides a measure of the amount of veiling glare present in the image.

The application of antireflection coatings to the lens surfaces has the effect of greatly reducing the brightness of all these effects that are due to interreflections between lens surfaces, and this is perhaps the principal reason why lenses are always coated. Lens coating, of course, has no effect on stray light arising from scattering inside the lens barrel, and this must be reduced by the judicious application of suitable baffles and by lining the lens barrel with black absorbent material. The sources of this kind of stray light can be located by placing the eye at a point in the image which should be dark and looking back into the lens. Stray light in a slide projector can be located by looking back into the projector from a point just outside the projected picture area.

F. Oblique Image Illumination

So far, we have confined our attention to a point on the axis of the lens. At points away from the axis the illumination falls off, and if the exit pupil of the lens were a small circle of constant diameter, the cos⁴ law would be accurately applicable in the image space. More usually the exit pupil is no longer a circle at higher obliquities, and in most lenses it changes its size, shape, and location as the obliquity increases. Put briefly, the image illumination is given by the product of object luminance, lens transmittance, cosine of the obliquity angle in the image space, and solid angle subtended by the exit pupil at the particular point in the field.

If the obliquity angle in the object space is under consideration, then we must take into account also the distortion in the image, in addition to the other factors of vignetting, cos⁴ law, and distortion of the entrance pupil. It can be shown[6] that in the absence of vignetting and pupil distortion, the combined effect of image distortion and the cos⁴ effect is given by

$$\frac{E_\phi}{E_0} = \frac{f^2 \sin \phi \cos \phi}{h'(\partial h'/\partial \phi)}, \qquad (5)$$

where ϕ is the obliquity angle in the object space, f the focal length of the lens, h' the image height, and $\partial h'/\partial \phi$ the rate of change of image height with obliquity angle ϕ (in radians) in the object space.

It is interesting to note that, with a distant object and in the absence of distortion, $h' = f \tan \phi$, so that $\partial h'/\partial \phi = f \sec^2 \phi$, and the ratio E_ϕ/E_0 becomes merely $\cos^4 \phi$, where ϕ is the obliquity angle in the object space.

If a lens is to have uniform illumination across the field (assuming uniform object luminance and no vignetting), then in the image space we must have an exit pupil distortion such that the solid angle subtended by the exit pupil at the particular field point is proportional to

---

[6] R. Kingslake, Illumination in optical images, *in* "Applied Optics and Optical Engineering" (R. Kingslake, ed.), Vol. 2, p. 214. Academic Press, New York, 1965.

TABLE V

Distortion Required for Uniform Field Illumination

Obliquity angle ϕ in object space (deg)	$h' = f \sin \phi$	Ideal image height ($f \tan \phi$)	Distortion (h' − ideal height)
10	0.174f	0.176f	−0.003f = −1.5%
20	0.342f	0.364f	−0.022f = −6.0%
30	0.500f	0.577f	−0.077f = −13.4%
40	0.643f	0.839f	−0.196f = −23.3%

sec ϕ', where ϕ' is the image-space obliquity. In the object space, we must have an amount of distortion such that the image height is given by $h' = f \sin \phi$, where ϕ is the object-space obliquity. In that case, $\partial h'/\partial \phi = f \cos \phi$, and Eq. (5) tells us that $E_\phi/E_0 = 1.0$, exactly. The required amount of distortion is surprisingly large (as shown in Table V).

CHAPTER 8

Projection Systems

I. A SELF-LUMINOUS OBJECT

A self-luminous object such as a cathode ray tube display can be projected onto a screen by means of a lens, but only at a considerable loss in brightness. Suppose B is the source luminance, m the image magnification, and N the F-number of the projection lens with transmittance t. Then if $m_p = 1$, the screen illuminance will be given by

$$E = \pi B t / 4 N^2 (1 + m)^2.$$

For example, if $\pi B = 5000$ fL, $t = 0.85$, $N = 2$ for an $f/2$ lens, and $m = 5$, then the screen illuminance is found to be $E = 7.4$ fc. If the screen is made of white paper, the image luminance will be about 7.1 fL a huge drop from the original tube luminance of 5000 fL. This can be helped by the use of a semispecular screen, but still the result is disappointingly low. This problem arises in projection television, where three tubes are often employed (one for each color), projected in register by high-aperture Schmidt mirror systems.

II. AN ILLUMINATED OBJECT

A. An Arc Projector

Professional motion picture projectors employ an arc source (either carbon or xenon) which is imaged into the film gate by a condenser lens, or more often by a concave elliptical mirror (Fig. 8.1), with the arc at one focus of the ellipse and the film gate at the other. The arc source may have an effective diameter of only 5 or 6 mm, and with a mirror magnification of five times, the image of the arc just fills the film gate having an area of 16 × 22 mm. The projection lens then forms an

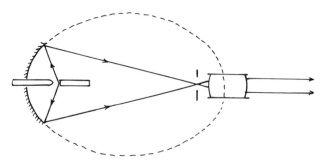

FIG. 8.1. An arc projector.

image on the screen which is a combination of the picture and the crater of the arc.

The magnification in the image formed by an elliptical mirror varies from zone to zone, with its magnitude along each ray equal to L'/L (Fig. 8.2). This leads to a large amount of coma in the image of the arc source, with the fortunate result that the film gate is uniformly illuminated even though the source itself may be somewhat irregular and variable. The mirror pickup angle is often about $\pm 70°$ and the imaging cone about $f/2.3$, which is easily contained in an $f/2$ projection lens.

Since $B' = tB$ (Section II of Chapter 7), the luminance of the projection lens aperture as seen from the screen is equal to the luminance of the arc source (except for losses). This may reach from 200 to 700 cd/mm^2 (20,000 to 70,000 stilb). In spite of the losses, and even though the rotating shutter admits only 50% of the light, it is possible to obtain a luminance of 10 to 15 fL on a 20-ft-wide screen in an ordinary theater. In outdoor theaters where the screen may be 60 ft. wide, the illumination is likely to be only about 1 or 2 fc.

B. An 8-mm Film Projector

In most modern 8-mm film projectors, a filament is mounted at one focus of an elliptical mirror and the film at the other. Because of the high cone angle of the imaging beam for use with an $f/1$ projection lens, the coma is very great, and including the effects of small random errors in the mirror itself, no sign of a filament image is seen on the screen. The mirror may be mounted inside a cylindrical glass enve-

II. AN ILLUMINATED OBJECT

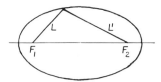

FIG. 8.2. Image magnification at an elliptic mirror.

lope, or may constitute the back of the envelope in the manner of a sealed beam automobile headlamp.

C. A SLIDE PROJECTOR

Slide projectors for home and lecture purposes, and many 16-mm film projectors, employ a filament lamp as source and a two-stage relay illumination system similar to the Köhler illumination in a microscope. A condenser lens forms an image of the lamp filament in the aperture of the projection lens, which in turn forms an image of the slide on the screen (Fig. 8.3). The slide must be placed close to the condenser so that it will be uniformly illuminated, and only the projection lens need be corrected for aberrations. As always, the illumination in the screen image will depend on the luminance of the light source and the angle subtended at the screen by the image of the filament in the projection lens aperture. If the projection lens is not filled with light, obviously the dark portions will contribute nothing to the screen illumination. Therefore, it is the aim of the projector designer to use as bright a source as possible and magnify it so that it will fill as much of the projection lens aperture as possible.

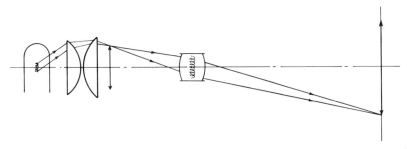

FIG. 8.3. A slide projector.

Unfortunately, these two requirements are somewhat contradictory, because the filament will have the highest luminance if it is made small and compact, which then calls for a high condenser magnification. In practice, the greatest condenser magnification that can be obtained in an ordinary slide projector is about 4×, which is often insufficient to fill the projection lens aperture completely. However, this is an advantage, because the spherical aberration of the condenser causes the filament image to wander sideways from point to point over the slide, and a little leeway leads to a more uniform illumination across the screen. Clearly, there would be no advantage in replacing the standard $f/3.5$ projection lens with an $f/2.8$ lens if the original lens is not filled with light.

It is worth noting that the beam of light emerging from the condenser consists of a uniformly illuminated central cone with an increasingly nonuniform and broad surrounding area, culminating in a sharp image of the filament at I (see Fig. 8.4). Thus the slide or film to be projected should be inserted into the central cone, and as the brightness of the cone increases toward its tip, the film should be inserted as far down the cone as possible without encroaching on the nonuniform surrounding area.

Recently some small and very bright lamps have been made by the addition of iodine inside a quartz envelope, which is run almost red hot. The iodine combines with the tungsten vapor from the hot filament, but eventually the tungsten iodide vapor comes into contact with the filament, where it is decomposed and deposits its tungsten back onto the filament and liberates the iodine. This cycle continues as long as the lamp is lit, thus preventing the deposition of tungsten on the inside of the envelope that occurs in the older type of lamp. The filament in these tungsten iodine lamps is often a single helical coil lying along the axis of the system, the lamp being surrounded by a deep elliptical reflector that is filled with light and acts as the "source" in the

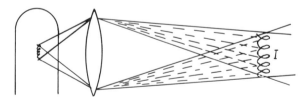

FIG. 8.4. The beam emerging from the condenser of a slide projector.

II. AN ILLUMINATED OBJECT

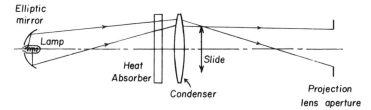

FIG. 8.5. Use of a tungsten iodine lamp in a slide projector.

projection system (Fig. 8.5). A very weak condenser lens is sufficient to image the mirror aperture into the projection lens aperture.

D. The Vuegraph, or Overhead Projector

A recent development, which is especially useful in classrooms or lecture halls, is the so-called *vuegraph,* or overhead projector (Fig. 8.6). In this instrument a small light source (often a tungsten iodine lamp) is imaged by a Fresnel condenser C into a projection lens L. Close to the condenser is a glass platen on which the lecturer can write or place a previously made transparency. Above the projection lens is a mirror at 45°, to project the image onto a vertical screen behind the lecturer's head. Rays from the top of the platen at A go to the screen at A' and from B to B', so that the image will be erect on the screen if it appears upright to the lecturer when facing the audience.

If the mirror is tilted above 45°, to raise the image on the screen for increased visibility to the audience, the image becomes keystoned, with a vanishing point at V where a line drawn at right angles to the axis of the beam strikes the screen. This form of distortion is commonly seen when a vuegraph is used, but most audiences accept it as inevitable.

There is a type of portable vuegraph in which the lamp S is mounted alongside the projection lens L (Fig. 8.7). A Fresnel lens and a plane mirror are placed immediately below the platen, to autocollimate the light and form an image of the lamp filament in the projection lens aperture. One result of this arrangement is that if the transparency, or the lecturer's pencil, is raised above the platen, a double image will be seen, one image caused by light coming down from the lamp and the other by light rising toward the projection lens.

8. PROJECTION SYSTEMS

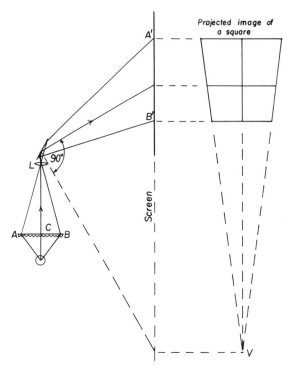

FIG. 8.6. A vuegraph, or overhead projector.

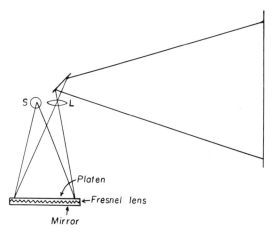

FIG. 8.7. A portable vuegraph.

II. AN ILLUMINATED OBJECT

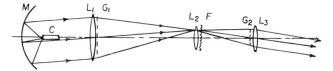

FIG. 8.8. Eidophor projection system (schematic).

E. Total Screen Lumens

One measure of the "efficiency" of a projector, by which two projectors may be compared, is the total number of lumens in the screen image from a slide of given size. This may be determined by dividing the screen image into a number of foot-square areas and measuring the illuminance in foot-candles at the center of each area. The simple sum of all these measures represents the total number of lumens falling on the screen because 1 fc is 1 lm/ft^2.

F. The Eidophor Projector

In the early 1950s, a unique television projector known as the Eidophor[1] was developed in Switzerland. Today it would be called a light valve because, in the absence of a signal, the field would be dark, and it would light up when a signal appeared. In this device a thin film of oil was scanned by a beam of electrons in a TV raster pattern, and as the intensity of the electron beam varied with the incoming signal, the oil film would display ridges and troughs following the television image.

To project this oil film on a screen, a theater arc lamp C and a mirror M were used, with a pair of coarse gratings in a schlieren arrangement (Fig. 8.8). A collimator lens L_1 focused the light source onto the oil film F, and the first grating G_1 was imaged onto the second grating G_2 by a field lens L_2 located close to the oil film. The two gratings were set so that in the absence of any signal the bars of one grating fell onto the spaces of the other. As soon as a signal appeared, some of the light would be refracted by the hills and valleys in the oil film, and so find its way past the second grating and on to the screen.

---

[1] E. Labin, "Eidophor theater television." *J. SMPTE* **54**, 393 (Apr. 1950).

III. PROJECTION SCREENS

A. OPAQUE

The simplest projection screen is, of course, a plain white diffusing surface, which will appear equally bright when viewed from any distance or at any angle. This is quite satisfactory provided the image is bright enough for comfortable viewing.

However, in many cases, it is desired to redistribute the light in such a way as to increase the screen luminance for some regions of the audience while reducing it for others. This is referred to as *screen gain*. For instance, a metallized or beaded screen has the effect of greatly increasing the luminance for viewers located close to the projector while reducing the luminance when the screen is viewed from a steeper angle.

A more recent procedure is to cover the screen with fine vertical grooves that scatter light laterally but not vertically, so that, whereas a fairly extended audience can view the screen comfortably, it would appear dim to anyone situated above or below the average audience level. A graphical plot of these screen types is shown in Fig. 8.9.

A curved screen. By far the brightest means of projecting images for observation is to mount the projector at or close to the center of curvature of a metallized concave spherical surface. The viewer then stands close to the projector, and the image appears intensely bright, even if sunlight is falling obliquely on the screen. The larger the screen

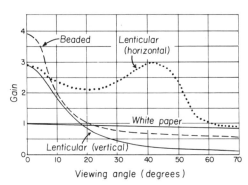

FIG. 8.9. Typical examples of screen gain.

the dimmer the image, but a large screen has the advantage of giving the viewer some latitude in placing his eyes to receive the maximum screen luminance.

B. Translucent

For many purposes, for instance in a microfilm or microfiche viewer, it is highly desirable to use a translucent screen with the projector on one side and the viewer on the other. Many problems arise, however, depending on the degree of diffusion in the screen material.

If the diffusion is small. If the amount of diffusion is small, as with slightly ground glass, the situation is indicated in Fig. 8.10a. The observer sees a bright "hot spot" surrounding the projection lens which moves around if he moves his head sideways. (Of course, if there were no diffusion, he would be looking directly at the projection lens aperture.)

The hot spot can be enlarged by the use of a field lens mounted in front of the translucent screen (Fig. 8.10b). A Fresnel lens works well, but the grooves must be narrow enough not to be visible to the observer. The field lens forms a real image of the projection lens, and if

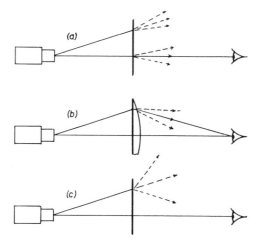

FIG. 8.10. Translucent projection screens.

the observer's eye is placed within the image, the screen will appear intensely bright, but if he moves his eye even slightly outside the limits of this exit pupil, the image will immediately disappear. The more diffusion that is used in the screen, the larger this exit pupil becomes, and the easier the system will be to use, but the dimmer the image will appear.

If the diffusion is large. If the degree of diffusion on the screen is sufficiently large, no field lens is needed, and the hot spot becomes large enough to cover the whole screen, but the image becomes relatively dim (Fig. 8.10c). This is generally the most satisfactory arrangement. However, ambient light now becomes a problem, because it is scattered back into the observer's eyes from the screen, and he soon finds that he must work in a darkened room. Placing a layer of neutral-density glass close to the screen often helps, because the image has to traverse the layer once whereas the ambient light has to traverse it twice. Home television receivers often have a neutral-density faceplate for this reason.

IV. STEREOSCOPIC PROJECTION

A stereoscopic pair of positive transparencies can be viewed by projection if some means is provided to ensure that each eye sees only its correct image, even though the two pictures are superposed on the screen.

In the colored anaglyph process, for example, one projected image is colored red and the other with the complementary blue–green. Similarly colored filters are worn by the viewer in the form of spectacles, so that the right eye sees only the right-hand image and the left eye only the left-hand image. Present-day stereo projection is performed with polarizing filters in front of the projection lenses, the viewers wearing similar polarizing filters in front of their eyes. By setting the Polaroids at 45° to the horizontal, it is immaterial which way round the filters are worn, but they must not be turned upside down or a pseudoscopic image will result. The screen must be metallized to prevent depolarization of the images.

It is not generally realized that stereoscopic projection is quite different from viewing stereo prints on paper. In projection the room is dark and only the light portions of the images are seen; on paper the background is light and only the dark lines are visible. In projecting a colored anaglyph, the background is colored red or green, and each image is viewed through the same colored filter. On paper, however, one image may be printed in green ink and viewed through a red filter to make it visible, and vice versa.

In stereo projection, the correct three-dimensional reconstruction is seen only if the observer is located at the center of perspective of the scene, somewhere around the middle of the audience. If the viewer is too close to the screen, the depth effect is diminished, and if he is too far from the screen, the third dimension is exaggerated. Some "3-D" movies have been produced for public viewing, but the result was satisfactory for only a small group of viewers near the center of perspective, and most people were not happy at having to wear cardboard spectacles while watching the movie.

V. CONTOUR PROJECTORS

A contour (or profile) projector is an instrument for projecting a magnified image of a small mechanical part upon a drawing of the part, to check its dimensions. Common magnifications are 5×, 10×, 20×, 50×, and 100×. The magnification ratio must be maintained exactly, and no distortion is permitted in the lenses. The optical system is invariably telecentric, with an aperture of $f/10$ or $f/16$, and a turret of lenses is used to vary the magnification. The virtue of a telecentric system is that the size of the projected image does not alter if the image is slightly out of focus (Fig. 8.11). As always in telecentric systems, the

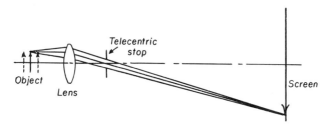

FIG. 8.11. A telecentric lens in a contour projector.

142 8. PROJECTION SYSTEMS

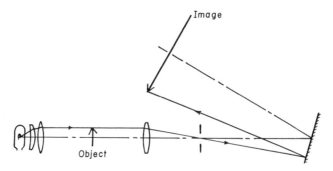

FIG. 8.12. A typical 5× contour projector.

diameter of the lens aperture must be greater than the diameter of the object. The condenser lens near the lamp, providing collimated light, must also be larger than the largest object to be projected.

The final image is generally projected on a ground glass screen, sometimes equipped with a Fresnel lens to brighten up the outer parts of the screen. As shown in Fig. 8.12, the projected image is upside down but correct as viewed by the operator standing behind the object.

A more elaborate contour projector was developed by the Eastman Kodak Company soon after World War II. It is shown diagrammatically in Fig. 8.13. A long symmetrical telecentric relay was mounted

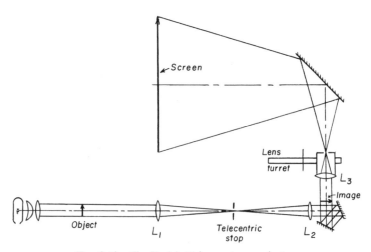

FIG. 8.13. The Kodak 14-in. contour projector.

behind the object, to provide a working distance of about 7 in. and erect the projected image on the screen. A turret of six projection lenses was provided, from 10× to 100×, and the working distance of these lenses could be quite small, as the relay provided the needed long working distance at the object holder. A telecentric stop was placed at the middle of the relay and another after each of the projection lenses. The screen diameter was 14 in.

CHAPTER 9

Plane Mirrors and Prisms

I. RIGHT- AND LEFT-HANDED IMAGES

If an object is placed in front of a plane mirror, each point in the object will be imaged, along the normal to the mirror, at the same distance behind the mirror as the object is in front. In this way a virtual image of the whole object is formed in the mirror. A side view of the situation is shown in Fig. 9.1. When you look at your image in a mirror, everything remains the same except that your image is reversed *along the line of sight*. Your left side is imaged at the left and your head is imaged at the top; the only difference is that the person in the mirror appears to be left-handed (if you are right-handed). We say that the *hand* of the image has been reversed by the mirror. When you look in the mirror of your car and watch another car coming up behind, you see that the driver of the other vehicle is sitting at the left side just as you are, and you do not notice anything peculiar until you look at his number plate, which you find is reversed, so that

| PGS-693 | appears as | ƐꝐᘒ-ꙅƆꟼ |.

In the printing industry, lead type must be in "looking-glass language" so that when it is inked and pressed on paper the lettering will come out correctly. Printers use the terms "right-reading" and "wrong-reading" to describe these two situations. In lithography, for example, a right-reading document is inked and rolled against a rubber roller, where it becomes wrong-reading. The image on the roller is then transferred by pressure onto a sheet of paper passing under the roller, thus a right-reading copy of the original document is produced.

When using a slide projector, the operator stands behind the projector to insert the slides. He first orients the slide so that it reads correctly as he sees it, and he must then turn it upside down, i.e., he must rotate it in its own plane through 180° before inserting it into the projector, because the lens in the projector will turn the image through another 180° on the screen. The image will then appear right-reading to him

I. RIGHT- AND LEFT-HANDED IMAGES

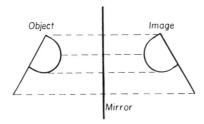

FIG. 9.1. Image formation by a plane mirror.

and to the audience on his side of the screen. However, if the screen is translucent and the audience is on the other side of it, they will see the image reversed left-for right. In that case, the operator must turn the slide upside down by rotating it about a horizontal axis in the plane of the slide before inserting it into the projector. This will invert the slide and also reverse it left-for-right as required. The lens will undo the inversion and the translucent screen will undo the reversal.

The early Photostat machines formed an image of a document on sensitive paper held inside the machine, and a mirror was mounted ahead of the lens to yield a right-reading image on the paper. This also happened to be convenient, as it permitted the document to be laid on a horizontal easel, as shown in Fig. 9.2.

To obtain a right-reading image when viewed in the same direction as the light is traveling, it is necessary to have an even number of mirrors; if the image is viewed toward the source of light, an odd number of mirrors is required. In the early Polaroid cameras the entering light formed a wrong-reading image on the photographic negative, which was jettisoned, but by diffusion transfer a right-reading

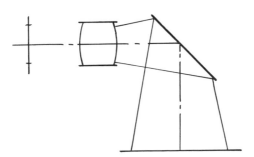

FIG. 9.2. Side view of a Photostat camera.

paper print was formed, so no mirrors were needed. In the later SX-70 system, the camera original was kept as the final print, and then a mirror was needed inside the camera to reverse the image. In the daguerreotype cameras of 1840, the camera original was kept, and a mirror should have been used to give a right-handed image. As this mirror was usually not provided, many early daguerreotype pictures were left-handed, but this was not noticed unless some lettering appeared in the picture or the viewer knew what the scene should have looked like. In Wolcott's mirror camera of 1842, the image was formed on the daguerreotype plate by means of a large concave mirror, and in this case there was no left-for-right reversal of the scene.

A. Plotting the Section of a Light Beam at an Inclined Mirror

The plotting of a section of a light beam at an inclined mirror is shown in Fig. 9.3. We assume that the lens has a circular aperture and the image projected by the lens has a square outline. We wish to plot the trace of the light beam as it strikes a plane mirror set at 45° to the lens axis, as shown in Fig. 9.3.

We first draw the side view of the beam, proceeding from the lens to the top of the image without the mirror in place (the beam is repre-

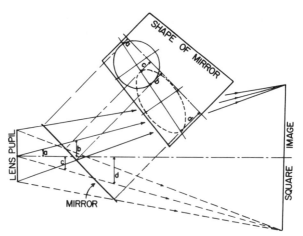

Fig. 9.3. The intercept of a light beam at a 45° mirror.

sented by its upper, principal, and lower rays). Because the beam is a circular cone, an intercept on a plane surface such as the mirror will be an ellipse. The side view of the mirror is also drawn in.

We now draw projection lines, perpendicular to the mirror out to one side, to form a picture representing the surface of the mirror, with center line marked. The projections from the points where the upper and lower rays intersect the mirror indicate the ends of the major axis of the ellipse, and there will be a row of these ellipses across the mirror for image points lying across the top of the square image. The extreme ellipses will have a major axis lying at distances a and b from the center line of the mirror, where a and b are the heights of the principal ray above the lens axis at the two ends of the ellipse (as read off the ray diagram). To determine the minor axis of the ellipse, we draw a line perpendicular to the major axis at its midpoint and mark off along it distances equal to the semidiameter of the cone of rays at the midpoint of the ellipse. Knowing the major and minor axes, we can plot the ellipse and trace it in position on the picture of the mirror.

We now repeat the whole process for a beam proceeding from the lens aperture to the bottom of the image. This gives us another ellipse, longer and narrower than before, and we notice that the major axis of this limiting ellipse, the ends of which are at c and d, lies in the same straight line as the major axis of the previous ellipse. We can now draw in the limiting size of the mirror, to contain all these ellipses and others lying between them.

The quality and flatness of the mirror surface requires attention. If it is flat within two or three Newton's rings over the area of each elliptical beam section, the definition of the image will be acceptable. However, if the mirror as a whole is wavy or twisted, the image will be somewhat distorted, although the definition may still be satisfactory. Mounting a thin plane mirror without flexure or other distortion is difficult and requires great care in planning the mounting. Often a three-point support is best. Alternatively, the mirror could be cemented at three points to a metal plate, which in turn is mounted in the instrument.

B. Procedure for Drawing an Ellipse

Knowing the lengths of the two semiaxes a and b of the ellipse, it is an easy matter to complete the curve. The radii of curvature at the ends

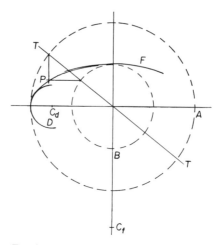

Fig. 9.4. Procedure for drawing an ellipse.

and sides of the ellipse are equal to b^2/a and a^2/b, respectively, and we can draw in these arcs immediately. The only problem is how to join them up. We can plot other points on the ellipse by drawing the exterior and interior auxiliary circles about the two axes and adding a transversal through the center, as shown in Fig. 9.4. Then the intersection of horizontal and vertical lines through the points where the transversal crosses the two circles is another point on the ellipse. Adding two or three such points between the arcs drawn previously will make it an easy matter to complete the curve.

II. ROTATING MIRRORS

If a mirror is rotated through an angle θ about an axis perpendicular to the plane of incidence, the reflected ray turns through twice as great an angle, namely, 2θ. However, if the axis of rotation is not perpendicular to the plane of incidence, some very complicated situations can arise.[1] These are best studied by regarding the mirror as being located at the

---

[1] G. F. Marshall, Scanning devices and systems, in "Applied Optics and Optical Engineering" (R. Kingslake and B. J. Thompson, eds.), Vol. 6, p. 203. Academic Press, New York, 1980.

II. ROTATING MIRRORS

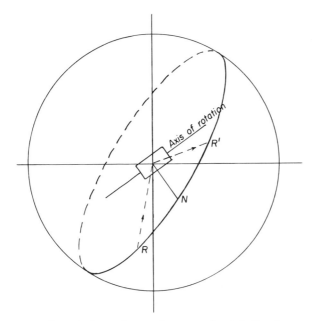

FIG. 9.5. A plane mirror at the center of an infinite sphere.

center of an infinite sphere (Fig. 9.5). Then the incident ray R, reflected ray R', and normal N are each represented by a point on the infinite sphere, and the plane of incidence becomes a great circle on the sphere joining the three points. The angles of incidence and reflection are projected as equal distances RN and NR' along the great circle. Now, if the mirror is rotated about some axis, the normal moves to a new position N' and the plane of incidence becomes a new great circle passing through R and N'. This enables the new reflected ray to be located.

A. THE SIDEROSTAT

One form of rotating mirror that has been used in astronomical instruments is the *siderostat*. In this instrument a plane mirror is rotated once in 24 hours about a polar axis, and it is set so that light from any celestial object is reflected along the polar axis itself. For example, the siderostat mirror can be mounted at the lower end of a

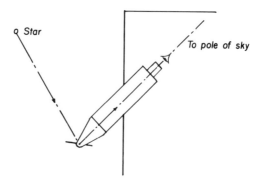

Fig. 9.6. A siderostatic telescope.

telescope, with the telescope tube pointing toward the pole of the sky and rotating once in 24 hours, carrying the plane mirror with it. This arrangement permits an easy and convenient location of the eyepiece inside a building, with a downward viewing position for the observer (see Fig. 9.6).

Of course, the siderostat can be mounted at the top of the telescope, as has been done in the large McMath solar telescope on Kitt Peak in Arizona (Fig. 9.7). In this case, only the siderostat mirror is rotated; the telescope itself is partly above ground and partly underground and the

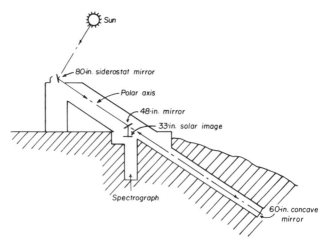

Fig. 9.7. The McMath solar telescope at Kitt Peak.

II. ROTATING MIRRORS

telescope axis points toward the pole of the sky at an elevation angle of 32°, which is the latitude of Kitt Peak.

B. The Coelostat

It should be noted that the image formed by a siderostat rotates once in 24 hours, which may be inconvenient, especially if a long exposure is involved. Consequently, a *coelostat* is preferred for many astronomical applications. In this instrument (pronounced "see-lo-stat"), suggested in 1895 by Lippmann, the polar axis lies in the plane of the mirror, which is rotated once in 48 hours (Fig. 9.8). Because the reflected ray from some celestial object may be steeply inclined or close to the horizon, a second mirror is required to direct the light into any desired direction. The coelostat yields a right-handed image.

C. The Heliostat

An early type of rotating mirror was called the *heliostat*. Several variations exist, one being suggested by S'Gravesand in 1742 and another by Foucault in 1869. A typical heliostat is indicated schematically in Fig. 9.9. The mirror C is mounted on a separate stand, where it can rotate about both vertical and horizontal axes. Axis *bd* is a polar

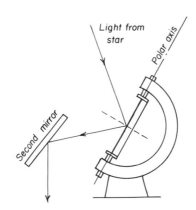

FIG. 9.8. Typical arrangement of a coelostat.

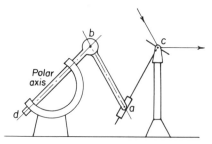

Fig. 9.9. A Foucault heliostat. (Line *ab* points toward the star, and *ba* = *bc*.)

axis rotating once in 24 hours; the arm *ab* points to the star being observed and the arm *bc* points toward the observer. The slider *a* rides along a rod held normal to the mirror, and *bc* equals *ba*.

D. Rotating Mirror Drums for Scanning

A rotating mirror drum (a "polygon") can be used to produce a straight-line scan. One arrangement is shown in Fig. 9.10. The mirror drum is mounted at the stop plane of a lens and the mirror facets are

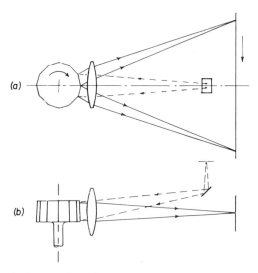

Fig. 9.10. Rotating mirror drum used to produce a line scan.

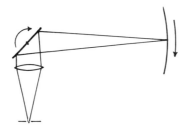

FIG. 9.11. A rotating mirror in a converging beam.

inclined to the rotation axis so that the reflected rays are always perpendicular to that axis. As each facet of the drum comes into place, a point of light sweeps along the focal plane of the lens. The incident light can be autocollimated by the lens, as shown in Fig. 9.10, or it may come from an independent collimated source. Of course, the lens must be specially designed to have a flat field when it is used in this particular way. A common requirement is that the length of the scan should be proportional to the rotation angle θ, requiring an "f-θ" lens.

If the rotating mirror comes after the lens, as shown in Fig. 9.11, then a cylindrical sweep is formed, which may be desirable under some circumstances.

III. MULTIPLE MIRROR SYSTEMS

A. A Two-Mirror System

A pair of mirrors mounted at a fixed angle θ to each other causes a ray to be deflected through an angle 2θ, three examples of which are shown in Fig. 9.12. In Fig. 9.12a, the angle θ is zero and the emerging ray is in the same direction as the incident ray. In Fig. 9.12b, the mirrors are at 45° and the ray is deflected through 90°. In Fig. 9.12c, the mirrors are at right angles and the reflected ray is parallel to the entering ray, but the direction of the light has been reversed. Note particularly that these deflection angles are independent of the slope of the particular ray that is entering the system, provided everything lies in one plane. A two-mirror system is therefore commonly referred to as a "constant deviation" system. For a ray entering from above or below the meridional plane, the pair of mirrors acts like a single

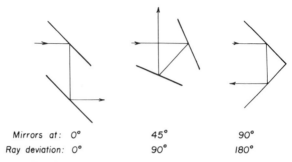

Mirrors at:	0°	45°	90°
Ray deviation:	0°	90°	180°

FIG. 9.12. Three examples of two-mirror systems.

mirror, so that a ray rising at, say, 10° from below the plane of the paper will continue to rise at 10° after reflection by the two mirrors.

B. The Triple Mirror

A set of three mirrors arranged like the corner of a cube has the property that it is a constant-deviation system for any arbitrary tilting of the system, with the reflected ray always returning parallel to the entering ray. The reason for this can be readily understood if an arbitrary tilt of the system is resolved into its three components, each parallel to one of the three coordinate axes where the pairs of perpendicular mirrors intersect. Each of these two-mirror systems is a constant-deviation arrangement for a rotation parallel to the line of intersection of the mirrors, so that, as a result, the final emerging ray is always parallel to the incident ray. Triple mirrors of this kind are often filled with glass, forming a cube-corner prism, which has the same constant-deviation properties. Sheets of cube-corner prisms are used along highways as markers, which glow brightly when light from an approaching car strikes them.

Of course, if the cube-corner reflector were made with absolutely plane surfaces and perfect right angles, it would not appear bright to the driver of the car because all the reflected light would return to the headlamp of the car from which it came. However, the ordinary methods of manufacture of glass or plastic panels filled with small cube-corners introduce just the right amount of spread in the returning beam to make the reflector visible to the driver of the car. Some perfect

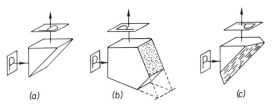

FIG. 9.13. Three prisms used to turn a beam through 90°.

cube-corner reflectors have been placed on the surface of the moon to reflect a laser beam back to earth; by this means the distance of the moon can be accurately determined by timing the passage of such a beam from earth to the moon and back again.

IV. REFLECTING PRISMS[2]

A *reflecting prism* is a piece of glass bounded by plane surfaces acting as mirrors, some of which may also transmit light. Typical examples of reflecting prisms used to deflect a beam of light by 90° are shown in Fig. 9.13. The image in each of these prisms is different. Case (a), an ordinary right-angled prism having only a single mirror, forms a left-handed inverted image, and if the object is a letter P, the image will be b. Case (b), a penta-prism, has two mirrors, leading to an erect image P. Case (c), an Amici prism with a roof edge in place of a mirror, gives an inverted right-handed image d.

The action of a roof prism is shown in Fig. 9.14, together with an end view of the roof. The roof divides the beam into two halves, one half entering the left side and crossing over to the right side, and the other half behaving conversely. The roof angle must be very accurately equal to 90° or signs of a double image will appear; the usual tolerance is a few seconds of arc. A parallel beam entering the prism perpendicular to the end face strikes the roof faces at an angle of incidence equal to 60°, so no metallizing of the roof faces is needed.

A familiar prism is that used at the upper end of a microscope to enable the eyepiece to be inclined at 45° when the tube is vertical, for ease in observation. This prism is shown in Fig. 9.15a. The prism

---

[2] R. E. Hopkins, Mirror and prism systems, *in* "Applied Optics and Optical Engineering" (R. Kingslake, ed.), Vol. 3, p. 269. Academic Press, New York, 1965.

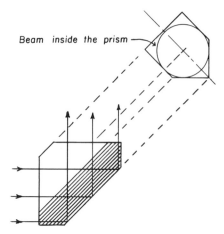

FIG. 9.14. A cylindrical beam inside a roof prism.

contains two reflecting surfaces, one of which is also a transmitting surface. The image formed by this prism is the same as the image that would be seen upon looking directly through the eyepiece if it were mounted in its usual place on the top of the tube. There is another form of prism that inverts the image formed by the microscope objective, and this prism involves a roof, as seen in Fig. 9.15b. These two prisms are actually the two halves of a roof Schmidt prism as shown in Fig. 9.22c.

A. The Optical Tunnel of a Prism

A prism is essentially a combination of a block of glass and one or more mirrors, and we can unfold a prism by reflecting it successively in

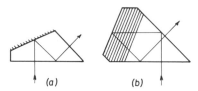

FIG. 9.15. Two prisms to deflect a beam by 45°, giving a right-handed image.

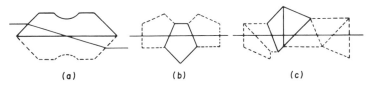

FIG. 9.16. Three optical tunnels: (a) Dove prism; (b) Penta prism; (c) Schmidt, or Pechan Z prism.

each of the mirror surfaces. It is best to draw the prism in its usual position and then reflect it outward in both directions, depending on what mirrors are present and how the axis of the beam is transmitted or reflected by each. Some examples of the so-called optical tunnel formed by unfolding a prism are shown in Fig. 9.16. Figure 9.16a shows an ordinary Dove prism along with its reflection in the hypotenuse face. This figure shows that the optical tunnel is a sloping block of glass and also that the entering and emerging rays are parallel. The result of unfolding a pentaprism is shown in Fig. 9.16b. The entering and emerging faces are parallel in this diagram, and the axis of the beam goes straight through the middle. If we hold a pentaprism in our hand and look into one end, we shall see the optical tunnel with all its angles and side extensions, just as in Fig. 9.16b. Finally, Fig. 9.16c shows the tunnel diagram of the Schmidt or Pechan prism. This prism consists of two parts with a narrow airgap between them, and each half is separately unfolded in the tunnel diagram. As before, the axis of the light beam passes straight down the middle of the diagram, and the entering and emerging end faces are parallel. It is easy to show that the length of the tunnel is equal to $2.5 + 3\sqrt{2}$ times the diameter of the beam, or $4.621D$. For the pentaprism the axial length is $2 + \sqrt{2}$, or 3.414 times the beam diameter.

B. Image Displacement Due to a Thick Glass Plate

Inserting a thick glass plate, or a reflecting prism, into a converging beam causes the image to be displaced along the axis, the extent of the displacement varying with the slope angle of the ray (Fig. 9.17). If the plate has a thickness t and refractive index N, the image will move

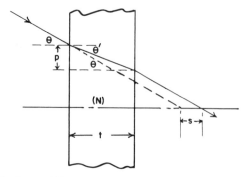

FIG. 9.17. Image shift caused by a parallel plate in a converging beam.

away from the lens by an amount

$$S = t\left(1 - \frac{\tan \theta'}{\tan \theta}\right) = \frac{t}{N}\left(N - \frac{\cos \theta}{\cos \theta'}\right),$$

where θ is the slope of a ray in air and θ' the slope of the ray inside the glass, so that $\sin \theta' = (\sin \theta)/N$. Clearly, the image displacement will increase as θ increases, leading to a considerable amount of overcorrected spherical aberration that must be corrected elsewhere in the system (it also leads to chromatic aberration). For paraxial rays, θ and θ' are so small that their cosines are equal to 1.0; hence the image shift for paraxial rays becomes

$$s = (t/N)(N - 1.)$$

For $N = 1.5$ (ordinary window glass), $s = t/3$. Thus the insertion of a plate of ordinary glass between a lens and an image shifts the image away from the lens by about one-third the thickness of the plate. Since the image magnification by a plane surface is unity, the size of the shifted image will be unchanged. This is a simple way to shift an image along the axis without any change in image size. However, if the desired longitudinal image shift is great, this procedure will introduce an excessive amount of aberration, and it is advisable to use some other method. One way is to use a pair of lenses with parallel light between them; moving one of the lenses will shift the image by the same distance as the lens has been moved. Another method is to use an afocal system with a magnification other than unity; shifting this system along the axis will cause the image to move at a much slower rate without any change in image size. The opposite problem, namely, to change the size of an image without moving it along the axis, can be solved by the use of a zoom lens of some kind.

C. Prism Angle Errors

If a prism angle is in error by a small amount ϵ, the reflected ray inside the prism is deviated by 2ϵ, and after emerging into the air this error becomes increased by a factor n, which is the refractive index of the glass. In the case of a roof prism, an error in the roof angle by ϵ results in an external error of 3ϵ in each half-beam, or 6ϵ altogether. If the roof prism is located close to the image, this image doubling might not be noticed, but if the roof prism is a long way from the image, it might generate a noticeable doubling, particularly because the eyepiece increases the angular doubling by its magnifying power, and the eye can detect details subtending about 2 arc-min. Thus if the roof angle error is 2 arc-sec, the prism doubling will be 12 sec, and if the eyepiece power is six times, this becomes 72 arc-sec, or just over 1 min. Although this would probably not be noticeable, it certainly indicates how accurately the angle of a roof must be produced.

D. The Prism Autocollimator[3]

A convenient instrument for detecting and measuring the errors of prism angles is the *autocollimator,* shown in Fig. 9.18. Suppose that a right-angled prism is mounted in front of an autocollimating telescope, with the telescope axis approximately perpendicular to the hypotenuse face of the prism. Three reflected images will be seen in the eyepiece. One of these, C, is faint and is caused by reflection at the hypotenuse face, whereas the other two, A and B, are caused by total reflection at the cathetal faces bounding the right angle. The separation between images A and B is six times the error of the right angle, and the lateral displacement of C from the line joining A and B is twice the "pyramid error," i.e., the angle between the hypotenuse face and the line of intersection of the cathetal faces. If the system were a true geometrical prism, this angle would be zero, and the presence of a

[3] B. K. Johnson, "Optics and Optical Instruments," p. 189. Dover, New York, 1960.

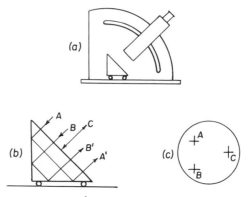

FIG. 9.18. The prism-testing autocollimator.

finite angle indicates that the so-called prism is really a section of a long thin pyramid. Many of the angles of a prism are either unimportant or can be cancelled by appropriate tilt or displacement of the prism in mounting, but nothing can be done to eliminate pyramid error, other than to introduce an equal and opposite pyramid error somewhere else in the system.

Another way to detect and measure the pyramid error in a right-angled prism is to mount the prism on its ground face on the table of a spectrometer, with the collimator and telescope nearly in line and the slit horizontal. Two images are seen, one is the Dove-prism image formed by light passing through the prism, and the other is formed by reflection from the outside of the hypotenuse face. When pyramid error is present, one of these images is raised while the other is depressed, and the vertical separation of these images can be read off a scale in the eyepiece of the telescope. If pyramid error is absent, the two images coincide and appear as a single image of the slit.

E. Tracking Images through Lenses and Mirrors

If a beam of light strikes one or more plane mirrors, we can follow its path by drawing a letter P on a card and moving it (in imagination) along the beam. Whenever an inverting lens is encountered, the card must be rotated through 180° in its own plane. When a mirror is encountered, the card must be tilted so as to lie in the plane of the mirror, and then allowed to continue in the reflected beam, so that now the back of the card leads the way. The next mirror again reverses the card, forming a right-handed image which may be erect or inverted. For example, between the two surfaces of a roof, the card, and therefore the image, is turned through a right angle by the first mirror, so that the image lies on its side and is left-handed. The second mirror of the roof completes the inversion and reverses the left-handedness.

Some imaging systems involving plane mirrors can be laid out by a paper-folding process. A strip of paper is cut with the lens aperture at one end, the image diameter at the other, and the length of the paper equal to the distance from lens to image. The paper strip is then folded as required by the problem, each fold representing the position and size of a plane mirror.

V. IMAGE ROTATORS

In some instruments, such as microfilm readers, it is necessary to be able to rotate an image about its midpoint. This is usually accomplished by inserting a rotator somewhere in the optical path. Five possible forms of an image rotator are shown in Fig. 9.19. The essential requirements are that the axes of the entering and emerging beams

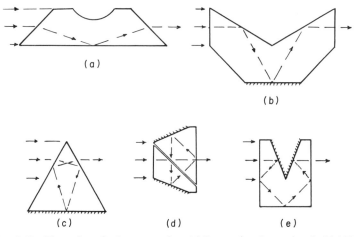

FIG. 9.19. Five prisms for image rotation: (a) Dove prism (one mirror); (b) Abbe K prism (three mirrors); (c) Taylor V prism (three mirrors); (d) Schmidt or Pechan Z prism (five mirrors); (e) Uppendahl M prism (five mirrors).

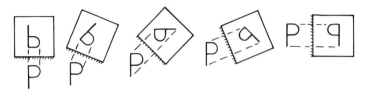

Fig. 9.20. Effect of rotating a Dove prism about the line of sight. (The reflecting hypotenuse is shown shaded.)

must lie in the same straight line, which constitutes the axis of rotation of the device, and there must be an odd number of reflecting surfaces.

The simplest rotator is the Dove prism, shown in Fig. 9.19a. The light is refracted at the sloping end face and bent down to the reflecting base, after which it is again refracted at the rear surface. The end angles can have any value, but they must be equal to each other; 45° is usual, but 40° prisms are common and more compact.

To understand why a Dove prism acts as a rotator, consider Fig. 9.20, in which an end view of the prism is shown in a series of angular positions. The object in each case is a letter P, and the reflecting base of the prism is indicated by shading. The figure shows that the image rotates in the same direction as the prism is turned, but at twice prism speed, so that when the prism has been turned through 90°, the image is completely inverted. Note that, because of the odd number of reflections, the image is left-handed, and another mirror must be included somewhere in the system to yield a final right-handed image. The same argument can be applied to the other rotators shown in Fig. 9.19, which have three or five reflecting surfaces. Note that prisms (a) and (c) have sloping end faces and consequently can be used only in parallel light; a converging beam becomes highly astigmatic if it is refracted at a sloping glass surface. Prism (b), the so-called Abbe or K prism, can be constructed from three plane mirrors instead of a solid glass prism; this is particularly useful if a large system is required where a prism would be impractical. The Schmidt, Pechan, or Z prism, shown in part (d), is the most used rotator because of its compactness and its perpendicular end faces that permit it to be used in a converging beam if required. The narrow air film makes mounting difficult. Prisms (b) and (e) must be composed of two parts cemented together, taking care that glass of the same refractive index is used for both parts. Surfaces shown shaded in Fig. 9.19 must be aluminized or silvered because reflection occurs there at angles of incidence less than the critical angle. Note that in many cases a surface is used for both reflection and refraction.

If we look through a Dove prism at an object that is rotating in a clockwise direction, its image in the prism will appear to rotate in a counterclockwise direction. Hence, if we wish to look end-on at a rotating object such as an airplane propeller or an electric fan, we must rotate the Dove prism in the same clockwise direction as the object to make it appear stationary.

A. Mounting a Rotator

It is assumed that the prism is mounted in a tube or ring which is free to rotate in a fixed bearing. The prism itself may be tilted or displaced within the tube, and the problem is how to adjust the prism position so that the midpoint of the image will remain stationary while the prism housing is rotated in its bearings.

To make this adjustment, we mount the prism, in its bearings, in a beam of parallel light between a collimator and a telescope. The telescope cross-wires must be previously adjusted to coincide with the image of the collimator pinhole. When the prism rotates, the image of the pinhole will, in general, describe a looped figure due to the following causes:

(*a*) the parallel beam may not be parallel to the axis of rotation of the prism housing;
(*b*) the prism may be tilted in its housing.

The first of these errors causes the image of the pinhole to describe a circle at *twice* prism speed, the second a circle *at* prism speed. The combination of both errors leads to a movement of the pinhole image having the general form shown in Fig. 9.21.

The adjustment procedure consists of two steps. First, the prism must be tilted in its mounting until the loop figure becomes a circle rotating at twice prism speed, so that in one revolution of the prism the pinhole image traverses a circle twice. Second, the pinhole itself is moved laterally until the circle shrinks to a point, which occurs when the parallel beam is parallel to the prism rotation axis.

This completes the adjustment if the prism is to be used in parallel light. However, if it is to be used in converging light, the lateral position

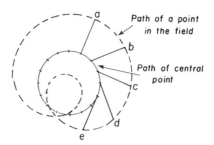

FIG. 9.21. Rotation of the image when the prism is tilted in its mount.

of the prism in its housing must be correctly adjusted, in addition to the two adjustments just described. To make this third adjustment, a well-centered lens is placed in contact with the collimator objective to produce a beam of converging light, and the observing telescope is replaced by a simple cross-wire and eyepiece, previously placed at the image of the pinhole with no prism in the beam. Any circular motion of the image at prism speed is now due to a lateral displacement of the prism in its holder, and this must be adjusted out until the circle shrinks to a point. Care must be taken, of course, that neither of the previous adjustments is upset by this third adjustment. Once the collimator has been set parallel to the axis of rotation of the prism housing, this adjustment need not be repeated if a whole run of identical prisms is being adjusted.

VI. PRISMATIC IMAGE ERECTORS

Prism systems are often used to erect an image, for example in a telescope. If the telescope is to be short, as in pocket binoculars, a prism erector is ideal. However, if the telescope must be long, as in a submarine periscope, then a lens erector must be used (see Section III of Chapter 12).

To design a prismatic image erector, we must remember that if a ray lies throughout its length in one plane and is reflected by an *even* number of mirrors, the image will be erect when the emerging ray is in the same direction as the incident ray, and it will be inverted if the emergent ray returns toward the source. In both cases the image will be right-handed. Therefore, to make a prism system that will invert an

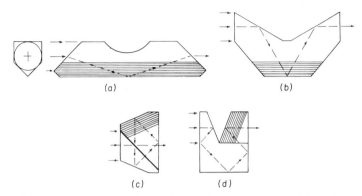

FIG. 9.22. Rotator prisms equipped with roof for image inversion: (a) Wirth; (b) roof Abbe; (c) roof Schmidt; (d) roof Uppendahl.

image, it is necessary to have an odd number of mirrors if the light continues on its way, as in a rotator. But the odd number of reflecting surfaces in a rotator yields a left-handed image, so it is necessary to replace one of the mirrors in a rotator with a roof. Any rotator with a roof instead of a mirror can be used to erect an image. In Fig. 9.22, four of the rotators shown in Fig. 9.19 are equipped with a roof, and the last two are used in several modern, high-priced binoculars.

It should be noted that a side view of a rotator and its equivalent with a roof are not identical, because a roof extends somewhat beyond the equivalent plane mirror, as indicated in Fig. 9.14.

Figure 9.23 shows three other roof prisms that have been used in small telescopes. They differ from those in Fig. 9.22 by having an offset between the entering and emerging beams. This may or may not be a convenience in the overall design of the instrument.

A. The Porro Prism

In the mid-1800s, I. Porro suggested that if a roof edge is made sufficiently large, so that only one of the two half-beams is used, it is no longer necessary to control the magnitude of the roof angle. His arrangement is shown in Fig. 9.24. The first prism inverts the image and reverses the direction of the light. The second prism merely reverses that direction of the light so that an eyepiece can be added. If the whole combination is now rotated through 90°, the roles of the

166 9. PLANE MIRRORS AND PRISMS

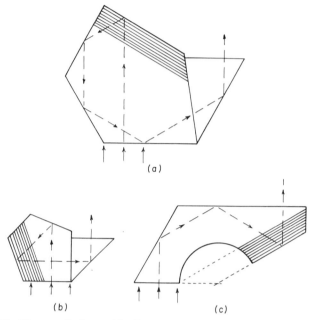

FIG. 9.23. Three roof prisms with offset beam: (a) Moller; (b) Hensoldt; (c) Leman.

prisms are reversed. At 45°, the first prism rotates the image through 90° and the second prism rotates it through an additional 90°, making 180° in all. Thus, as the combination is rotated, the action of the prisms continually changes, but the result is the same.

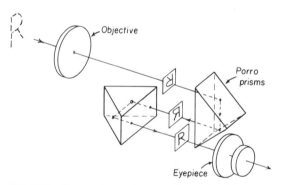

FIG. 9.24. Porro prism system used in a prism binocular.

As each prism is a rotator, it is necessary to mount the two halves together with high accuracy or the final image will lean. A small image lean would be insignificant in a monocular telescope, but in a binocular it is essential that the two eyes see identical images. Consequently, it is customary to ensure that each side of a binocular causes perfect image erection with no visible lean.

In large prism binoculars, for military purposes or bird watching, the two objective lenses are generally set further apart than the eyepieces to increase the stereoscopic effect. However, in small pocket binoculars, the objectives are often closer together than the eyepieces. Some modern binoculars equiped with roof prisms consist of two parallel tubes, which fold up extremely compactly to go in the pocket.

B. The Daubresse Prism

We have seen that if a prism reverses the direction of the light, an even number of mirrors leads to a rotator, as in one-half of a Porro prism. An interesting and little known prism that does not cause the image to rotate when the prism is turned is the Daubresse prism, shown in Fig. 9.25. It can be constructed from a right-angle prism with another right-angle prism cemented to each of its cathetal faces. Thus the prism contains three mirrors, and although it reverses the direction of the light, it does not act as a rotator. It turns the image through 45° so that it lies on its side, and it makes the image left-handed. A second similar prism reverses the left-handedness and turns the image through another 45°, thus yielding an erect or an inverted image depending on how the two prisms are connected. A Daubresse prism pair has been used in a large prism binocular in which the two tubes are rigidly connected and only the eyepieces and their half-prisms are movable to

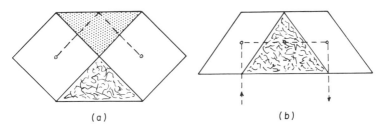

Fig. 9.25. The Daubresse prism, plan and side elevation.

168   9. PLANE MIRRORS AND PRISMS

alter the interocular separation. This procedure is impossible with an ordinary Porro prism pair.

C. THE PANORAMIC TELESCOPE

An interesting example of a telescope with prism erection is the familiar military panoramic telescope used to scan the entire horizon from a fixed position, often in a trench below ground level. The general arrangement is shown in Fig. 9.26. The head prism A can be rotated about a vertical axis to scan the horizon, but when this is done, the image rotates at prism speed. This rotation is compensated for by the use of a Dove prism B immediately below the head prism, acting as a derotator. The Dove prism must be rotated at half the speed of the head prism; this is accomplished by a type of differential gear similar to that in the back axle of an automobile. As there are now two mirrors in the system, it is necessary to add two more; this is done by the use of either a pentaprism or an Amici roof prism at D. In assembly, the

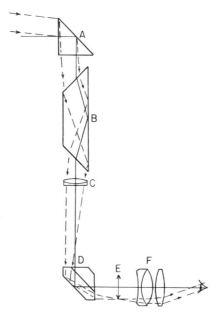

FIG. 9.26. The panoramic telescope.

VI. PRISMATIC IMAGE ERECTORS

Dove prism is rotated until the image is erect, after which the gears are meshed and the image remains erect permanently.

Figure 9.26 shows a typical arrangement, with the addition of a small $3\tfrac{1}{2}\times$ telescope to magnify the image. The true field at $\pm 5°$ is shown by dashed lines, the apparent field being $\pm 17°$ at the eye. The Dove derotator B is mounted ahead of the objective lens because it must be used in parallel light. Because angles will be read off the two rotations in this instrument, great care must be taken to mount the Dove prism accurately, so that the point in the middle of the field remains fixed.

CHAPTER 10

The Eye as an Optical Instrument

I. DIMENSIONS

The human eye (Fig. 10.1) is similar to a little camera, complete with lens, automatic focusing, automatic diaphragm control, a sensitive receiving surface curved to fit the curved field of the lens, and a full range of color sensitivity. The resolving power of the eye is, for most people, about 2 arc-min. or 1 in 1700. This means that under favorable circumstances most people might expect to see the millimeter markings on a ruler at a distance of about $5\frac{1}{2}$ ft.

The combination of eye and brain has many remarkable properties. In bright light the sensitivity of the system to light becomes automatically reduced so you will not be dazzled; this is known as *light adaptation*. On the other hand, after spending some time in a dark room, the sensitivity of the system greatly improves, so that after a few minutes many things that were invisible at first can now be seen. This is known as *dark adaptation*.

Another automatic visual compensation is concerned with image motion across the retina. If we move our eyes or head rapidly from side to side, the image of surrounding objects sweeps across the retina, yet we detect no motion of the objects themselves. There is an automatic feedback mechanism which tells the brain that the objects are actually stationary even though the images are moving. We can upset this mechanism by looking through a lens; in this case the objects do appear to move when we turn our head sideways.

The dimensions of the visual system have been determined by many observers. According to Gullstrand, the dimensions of the average relaxed eye are as follows:

focal lengths: anterior 17.1 mm; posterior 22.8 mm;
principal points: anterior 1.35 mm behind vertex;
                    posterior 1.60 mm behind vertex;
lens power: 58.6 diopters (cornea 43 diopters; lens 15.6 diopters).

I. DIMENSIONS

FIG. 10.1. Longitudinal section of human eye.

A. The Pupil Diameter of the Eye

One of the obvious involuntary properties of our eyes is the automatic variation in pupil diameter depending on the amount of light entering the eye. Assuming a very wide angular field of view with a definite luminance, many observers have plotted a graph of pupil diameter versus field luminance, as shown in Fig. 10.2. Of course, if the field is narrow and seen against a dark background, we would not expect the pupil diameter to shrink by the amount shown.

As indicated in Fig. 10.2, the largest and smallest average pupils are about 8 mm and 2 mm in diameter, respectively, and represent an area ratio of about 16 to 1. This by no means compensates for the million-to-one ratio of the brightest sunlight to the darkest starlight to which the eye is sensitive.

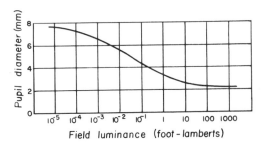

FIG. 10.2. Pupil diameter versus field luminance.

II. THE PROPERTIES OF VISION

A. The Resolving Power of the Eye

Many factors affect the ability of the eye to resolve fine details in an observed object. The retina itself has a granular structure, with each retinal element subtending an angle of about 30 arc-sec at the eye lens, so to observe resolution between adjacent details, we would expect that one retinal element must be unexposed between two stimulated elements, resulting in the well-known resolving limit of about 1 arc-min, which is 1 part in 3440, or 1 mm at about 11 ft. Most people can achieve this resolution only under the most favorable conditions, if at all.

Experiments show that the resolving power of the eye also depends on the object luminance and pupil diameter, as well as object contrast. At small pupil diameters the lens system of the eye is nearly diffraction-limited, but at large diameters it has a significant amount of aberration, both spherical and chromatic, which tends to reduce the resolution at large pupil diameters. Finally, experiments have shown that the acuity of vision varies with the distance of the object, owing to changes in the shape of the crystalline lens during accommodation. There is a great need for a standardized set of viewing conditions for any measurement of visual resolution, including the size of the resolution chart, the size and luminance of the surrounding area, and the distance of the observer from the test chart.

To illustrate the effect of some of these variables, a series of measurements by Fabry and Arnulf[1] on a single observer is reproduced in Fig. 10.3. A small test object consisting of equally wide lines and spaces was observed against a dark background, and the distance of the observer was increased until he could no longer see the separate lines on the target. The luminance of the bright lines was varied, as was the contrast between the bright and dark lines. The size of the pupil was varied artificially by letting the observer view the target through a series of holes in a thin metal plate. Under these specific conditions, the observed resolution of a single observer at high contrast is shown

---

[1] C. Fabry and A. Arnulf, "La Vision dans les Instruments." Editions de la Revue d'Optique, Paris, 1937.

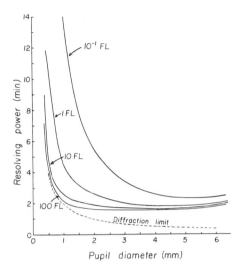

FIG. 10.3. Resolving power of the eye.

graphically in Fig. 10.3. If the experimental conditions had been changed—for instance, if the test object had been observed against a bright background—everything would be different because the pupil diameter would then have depended on the luminance of the background.

The bottom of the graph in Fig. 10.3 shows in dashed lines what the resolution would have been if the eye were diffraction-limited. It is clear that the eye is nearly perfect for a pupil diameter of 1 mm or less, but it is very far from perfect at higher pupil diameters. When the contrast in the target is lowered, all the curves move upward and the resolving power rapidly diminishes.

In very dim light, our ability to see an object at all drops rapidly. This explains the so-called night glass effect, where the use of a 7× binocular with a large exit pupil helps enormously in detecting distant objects under twilight conditions. Although the object actually appears dimmer when seen through the binoculars, the fact that it appears larger makes all the difference.

It is essential to distinguish between seeing and resolving. The eye can *see* a star subtending an angle of much less than an arc-sec, but it cannot *resolve* two stars separated by such a small angle.

B. Critical Flicker Frequency

If an observed illuminated image is switched on and off rapidly, a sensation of flicker is produced. However, if the flicker frequency is sufficiently high, the successive images merge into one another and all flicker vanishes. The critical flicker frequency depends on the scene brightness and type of flicker. The following tabulation gives the results of one determination of the critical frequency limit, for the case of a square-wave flicker with equal intervals of light and darkness.

Luminance		10^{-3}	10^{-2}	10^{-1}	1	10	10^2
	fL						
	Nits	0.0034	0.0343	0.343	3.43	34.3	343
Critical frequency/sec		13	19	26	34	43	51

The figures in this tabulation show that the audience in a movie theater will not notice the intermittent picture cycle provided the flicker frequency is at least 48 flashes/sec, but they would complain very much if the frequency were 24 flashes/sec. The projector manufacturers have had to insert a *flicker blade* into the rotating shutter of a projector to raise the flicker frequency from 24 to 48 flashes/sec, although the frame rate of the film is only 24 pictures/sec. Even this frequency would not be acceptable if the screen luminance were to exceed 100 fL, but there is little likelihood of this occurring.

C. Spectral Sensitivity of the Eye

Many determinations have been made of the response of the eye to monochromatic light of differing wavelengths (colors). Assuming that the radiant energy in the beam is held constant and only the wavelength is changed, the response of the average eye plotted versus a linear scale of wavelength is shown in Fig. 10.4.

Two curves are shown in Fig. 10.4. Curve (a) represents the scotopic sensitivity of the eye. It is plotted by measuring the least amount of energy needed in a monochromatic beam in order to be just visible to a dark-adapted eye. The peak of this curve falls in the blue–green region at about 5100 Å, and it will be noticed that red light is quite invisible

II. THE PROPERTIES OF VISION

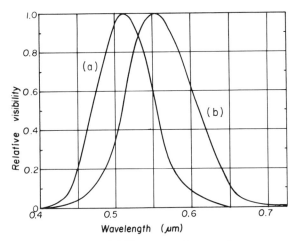

FIG. 10.4. Visibility curves of a normal eye.

under these conditions. Indeed, at sufficiently low levels of illumination, our color sense virtually disappears, and "all cats are grey in the dark."

Curve (b) in Fig. 10.4 is called the photopic curve, and it represents the response of the eye at everyday levels of illumination, with a peak at about 5550 Å. It is plotted by a process of heterochromatic photometry, in which the luminance component of a monochromatic colored area is matched visually against an adjacent white area of variable luminance. This type of observation requires much practice to obtain consistent results, and it is indeed a philosophical question as to whether it is possible in principle to make such a match. At fairly low levels of luminance the response of the eye falls between these two curves. The loss of color vision at low luminance levels is known as the *Purkinje effect*.

D. Apparent Brightness of an Object

The apparent brightness of an object, as seen by the eye, depends on several factors, principally the retinal illumination and the state of adaptation of the eye. The retinal illumination, in turn, depends on the luminance of the object and the diameter of the eye pupil. The relative aperture of the eye-lens system varies from $f/2$ to $f/8$, depending on the average luminance of the surroundings. For all these reasons, psychophysicists take care to distinguish between the physical *luminance* of an object and the *brightness* of that object as seen by the eye.

E. Astronomical Magnitudes

The relative visibility of stars and other astronomical objects is expressed in an arbitrary scale of "magnitudes." It has long been established that a change in apparent stellar brightness of 2.51 times should be regarded as one unit of magnitude. This number is the antilog of 0.4, so that five magnitudes is equal to a brightness ratio of 100 times. There are some 14 stars that have been assigned a magnitude of 1.0 (e.g., Aldebaran), and the eye can see stars as faint as the sixth magnitude, at which some 2000 stars have been recognized.

III. SPECTACLE LENSES

In the normal emmetropic eye, the image formed by the eye lens is sharply focused on the retina, and as the object moves closer to the eye, the power of the lens increases automatically to maintain sharp focus. This process is entirely involuntary and is known as *accommodation*. It is due to a change in the shape and power of the crystalline lens by muscular action. As we get older, the crystalline lens becomes stiffer and the available range of accommodation diminishes. This effect differs greatly from one individual to another, and Fig. 10.5 shows the range of measured accommodation for several thousand people.

These, and indeed most of the refractive properties of the eye, are expressed in diopters, the diopter being the reciprocal of the meter. Thus, infinity is 0 diopters, 50 cm is 2 diopters, 10 cm is 10 diopters, and so on. Hence, the 14-diopters range of accommodation of an average 10-year-old person with normal vision would run from about 0 to 14 diopters, i.e., from infinity to 7 cm. If he were near-sighted, his range of accommodation might run from, say, 2 to 16 diopters, i.e., from 50 to 6 cm, and so on. The two limits of a person's range of accommodation are called his *far point* and *near point*, respectively. Of course, vision does not end abruptly at the near and far points, but it

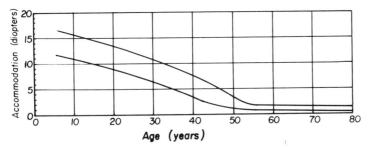

FIG. 10.5. Drop in accommodation with advancing years.

gradually deteriorates, and the person requires larger print to be able to read comfortably beyond his normal distance limits. However, these limits become fairly definite when we try to read very small print.

If a person's far point is closer than infinity, we say that he is near-sighted, or *myopic,* by so many diopters. To move the far point back to infinity requires the use of a negative spectacle lens of the necessary power. Thus, if his far point is 50 cm from the eye, a -2 diopter lens is required. The error is caused by the eye lens being too strong by 2 diopters, and adding a negative spectacle lens will weaken it to the normal emetropic condition.

On the other hand, in some eyes the lens is too weak. This is seldom discovered in childhood because the available range of accommodation is great enough to overcome the error unconsciously. Far-sightedness, or *hyperopia,* sometimes becomes evident at age 20 or later by the vision becoming blurred, even for objects at a great distance, when the person is tired, and a conscious effort is required to bring them into focus. The addition of a suitable positive spectacle lens brings everything into focus once more. Thus, for example, if a person's near point is at 10 cm and he has 12 diopters of accommodation, his far point will be at 2 diopters "beyond infinity," and a $+2$ diopter spectacle lens brings the far point back to infinity and moves the near point from 10 to 12 diopters (10 to 8 cm) from his eye.

Problems in accommodation and refractive errors can be solved by the use of a scale of distances in diopters, as shown in the following tabulation.

Diopters	0	1	2	3	4	5	6	7	8	9	10	11	12	13	14
Distance (cm)	∞	100	50	33	25	20	17	14	12	11	10	9	8	7½	7

The line drawn over the diopter scale indicates a myopic person with 10 diopters of accommodation; his far point is at 2 diopters and his near point 12 diopters. Because he has 2 diopters of myopia, he needs a -2 diopter spectacle lens. This lens has the property of moving the entire line by 2 diopters to the left, so that it would now run from 0 to 10 diopters. If he were to wear a $+2$ diopter eyeglass lens, the line would move by 2 diopters to the right, i.e., from 4 to 14 diopters (25 to $7\frac{1}{2}$ cm). On the other hand, for a hyperopic person having 10 diopters of accommodation and 2 diopters of hyperopia, the line extends to the left beyond the infinity point, and he needs to wear a $+2$ diopter lens to bring his vision back to normal.

When a person becomes so old that his range of accommodation is seriously reduced, he needs to wear eyeglass lenses for reading, although he may not need them for distant vision. Suppose he has 2 diopters of accommodation and 1 diopter of hyperopia; he can now see sharply from infinity to 1 m, but this is not close enough for comfortable reading. By wearing a $+2$ lens, his accommodation range is moved to the right, and he can now see clearly from 1 to 3 diopters, i.e., from 100 to 33 cm. He would probably prefer a somewhat stronger lens for close work, say $+3D$, in which case his range of sharp definition would extend from 2 to 4 diopters (50 to 25 cm).

A. Multifocal Lenses

Today, most old people wear multifocal (bifocal or trifocal) lenses in which selected areas of the lens have additional power for reading. Two diopters (2D) represents a convenient addition in the lower half of a bifocal, or 1 diopter addition in each segment of a trifocal lens. The upper area of the lens contains whatever power the person needs to see clearly at infinity, and this may be positive, negative, or even zero power.

Eyeglass lenses are generally manufactured with a fixed power on one surface and whatever power is needed on the other. A person requiring $+1D$ for distant vision and $+3D$ for reading might have a $-5D$ curve on the rear face of his lenses with a $+6D$ upper and a $+8D$ lower segment on the front face. A person with, say, 5 diopters of myopia might have a $+1D$ curve on the front and a $-6D$ curve on the back of his eyeglass lenses.

Some progress is being made in the development of eyeglass lenses having a progressively increasing power from top to bottom so that they act as a continuously variable multifocal lens. It seems that the decentered aspheric surface necessary for this introduces a large amount of astigmatism which is difficult, if not impossible, to correct.

B. Toric Lenses

Many people suffer from some degree of astigmatism in the middle of their visual field, mainly due to the pressure of the eyelid on the soft cornea. The result is that vertical and horizontal lines cannot be focused simultaneously. The cure for this refractive error is to use eyeglass lenses that have different powers in different orientations, with the meridians of maximum and minimum power being mutually perpendicular. At one time this was achieved by grinding a spherical surface on one face of the lens and a cylindrical surface on the other, but today a toric surface is ground on one face and a spherical (or bifocal) surface on the other.

A *toric* is a surface with different curvatures in perpendicular meridians, examples being the bowl of a teaspoon or the rim of an automobile tire. Thus, if a person has 0.5 diopters of astigmatism and no other error, his eyeglass lens may have a 5-diopter surface in front and a toric surface on the back with a power of -5 diopters in one meridian and either $-5\frac{1}{2}$ or $-4\frac{1}{2}$ diopters of power in the other meridian, according to his needs. Most people are very sensitive to even small amounts of astigmatism, and the powers and directions of the two principal meridians must be determined with great care.

The personnel of the spectacle industry include the *optometrist,* who measures the errors of a person's vision, and the *optician,* who makes the eyeglasses and mounts them in a suitable frame, in accordance with the optometrist's prescription. There is also the *ophthalmologist,* a medical doctor qualified to perform eye surgery, who commonly measures errors in vision also. The old term *oculist* has no particular connotation.

The capabilities of the eyeglass industry are truly remarkable. Where else would you find anyone capable of producing a pair of thin, light-weight lenses, cut to shape and mounted in a frame, with bifocal or trifocal powers on one side of each lens and possibly differing toric surfaces on the other side, and make the whole apparatus in a few days and sell it at a profit for under $100? Injection-molded plastic lenses are becoming increasingly popular in eyeglasses; they are light in weight and surprisingly robust. For those who can wear them, hard and soft contact lenses are now available; these rest in contact with the cornea under the eyelids and require no frame to hold them.

Removal of the crystalline lens to cure cataract causes the eye to lose some 16 diopters of power, and a very strong eyeglass lens is usually required. Of course, if the patient is normally near-sighted, his post-cataract spectacles would not have to be so strong, and some highly myopic patients find that they can see better after surgery than before. Contact lenses are particularly beneficial for after-cataract patients. Recently, some success has been achieved by implanting a plastic lens into the eye itself to replace the eye lens that was removed in the cataract operation.

IV. STEREOSCOPIC VISION

When we look at some object with both eyes open, each retina receives a slightly different image of the object, and the brain combines these two images to yield a sensation of depth. This effect is incredibly sensitive, and the slightest difference between the two images results in a depth sensation. Careful measurements have shown that the *stereo acuity* of the eyes is about 10 arc-sec, so that objects subtending 1 in 20,000 at the eyes produce some depth sensation. If the interocular separation is, say, 65 mm, the furthest object capable of depth sensation is at a distance of $20,000 \times .065 = 1300$ m. Of course, this limit can be increased by artificially increasing the stereoscopic base, as with a stereoscopic rangefinder.

A peculiar effect is observed if a close object is observed through a pair of prismatic rhombs placed in front of the eyes so as to increase or reduce the effective separation of the eyes. If the rhombs are turned outward so as to increase the effective interocular separation, near objects appear to be too small, an effect known as *dwarfism*. The reason is that the convergence of the eyes must be increased as a result of the increased interocular separation, so our brain tells us that the object is closer than it really is. Because the angular subtense of the object is not changed by the rhombs, we get the feeling that the object must be smaller than it really is. The converse is equally striking, and if the rhombs are turned inward so as to reduce the effective interocular separation, near objects appear larger than real, an effect known as *giantism*.

The images seen by our two eyes can be recorded photographically by the use of a stereo camera equipped with two lenses spaced apart, forming two slightly dissimilar images on the film. After printing, and rotating each print through 180° to erect the images, the two pictures are presented to the two eyes by means of a stereoscope. Ideally, the separation of the camera and stereoscope lenses should be equal to the observer's interocular separation (nominally $2\frac{1}{2}$ in.) and the focal lengths of the camera and stereoscope lenses should be identical. However, most practical cameras and stereoscopes depart somewhat from these ideal conditions, resulting in some degree of three-dimensional distortion. This is often considered desirable if it leads to a moderate increase in the depth sensation.

Some stereoscopic periscopes have been constructed. Two objective lenses are spaced apart and the two beams are polarized perpendicularly to each other. The two beams are then combined into a single relay system that goes right through to the eyepiece end of the instrument. At this point, a beam splitter divides the beam into two parts for the two eyes, and by the use of polarizing filters, each eye is made to see only its appropriate original image. Some light is lost by this process because a pair of parallel polarizers transmits less than 50% of the light, but the saving in having only one relay system for both eyes is considerable.

CHAPTER 11

Magnifying Instruments

I. THE SIMPLE MAGNIFIER

The apparent size of anything seen by the eye is expressed by the angle it subtends at the eye. Consequently, any device that increases angular subtense acts as a magnifier and makes objects look larger. The magnifying power (MP) of such a device is given by the ratio

$$MP = \frac{\text{angular subtense of the image seen through the instrument}}{\text{angular subtense of the object viewed directly}}$$

The denominator of this fraction needs some clarification, because merely bringing an object closer to the eye makes it appear larger. A myopic observer can often see a small object easily by simply removing his spectacles and bringing the object close to his eye.

For a simple magnifier or microscope, where an object can be placed at any desired distance from the eye, it is necessary to establish a conventional "nearest distance of distinct vision" at which the object is assumed to be viewed directly. This distance V is generally taken to be 10 in., or 250 mm.

We can now set up a general relation between the distances of object, image, and lens from the eye and obtain formulas for the magnifying power of the system. In Fig. 11.1, we see that if the object to be magnified is placed at a distance a beyond a simple thin magnifying lens and if the image seen by the eye is at a distance b from the lens, then if the eye is at a distance E from the lens, the magnifying power will be given by

$$MP = \frac{\text{image subtense}}{\text{object subtense}} = \frac{h'/(E + b)}{h/V} = \frac{mV}{E + b}, \qquad (1)$$

where m is the image magnification $h'/h = b/a$. It is essential to distinguish between magnification and magnifying power. *Magnification* is the simple ratio of image size to object size. *Magnifying power* is the ratio of image subtense to object subtense (at the conventional

I. THE SIMPLE MAGNIFIER

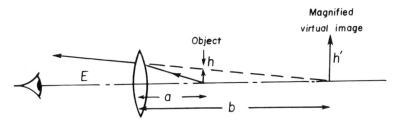

FIG. 11.1. Magnifying power of a lens.

distance V). Magnifying power thus involves object and image distances as well as their sizes.

It was shown in Section II of Chapter 4 that $1/a - 1/b = 1/f'$. Hence, solving for m gives

$$m = f'/(f' - a) \quad \text{or} \quad m = (b + f')/f'.$$

In the ordinary use of a magnifier, to make a small object look larger, the image magnification is not known, so we eliminate it, giving the two alternative expressions:

if the object distance a is known, $\quad \text{MP} = \dfrac{Vf'}{af' + E(f' - a)};\quad$ (2)

if the image distance b is known, $\quad \text{MP} = \dfrac{V(b + f')}{f'(b + E)}.\quad$ (3)

Several special cases may now be considered.

If the Eye Is Close to the Lens. This is by far the most common situation. Here, $E = 0$, $b = \infty$, and the object is at the focus of the lens, so $a = f'$. Thus, Eq. (11.2) tells us that

$$\text{MP} = \frac{Vf'}{f'^2} = \frac{V}{f'}.$$

Thus the magnifying power of a simple lens used in this way is merely equal to 10 in. divided by the focal length of the lens. An eye loupe having a focal length of 3 in. is labeled 3.3×, and a high-power pocket magnifier of 1-in. focal length is labeled 10×. Microscope eyepieces are also labeled by this system.

A slight gain in magnifying power can be obtained by accommodating the eye to a distance V while using the lens. Then $E = 0$ and $b = V$, so that by Eq. (2),

$$\text{MP} = \frac{V(V+f')}{f'V} = 1 + \frac{V}{f'}.$$

The magnifying power has thus been increased by 1.0, which is significant at very low powers but not at a magnifying power of 10.

If the Lens Is Distant from the Eye. We sometimes use a magnifier at some distance from the eye and move the object toward or away from the lens to give a desired magnifying action. Suppose that initially the object is touching the lens and $m = 1.0$. Equation (1) tells us that the magnifying power is then equal to V/E, which could be less than 1.0 if the distance of the lens from the eye is large. We could observe the object more easily by merely bringing it closer to the eye, without any lens at all!

Now suppose that we move the object slowly away from the lens, keeping the lens at a fixed distance from the eye. The image increases in size until it fills the whole lens, after which it reappears upside down, and gradually becomes smaller as the object is moved still further from the lens.

For example, if $E = 20$ in. and $f' = 2$ in., Eq. (2) tells us that the magnifying power will vary with a, in the manner shown in the following tabulation.

Object distance a (in.)	0	0.5	1.0	1.5	2.0	2.222	2.5	3.0	3.5 ...
MP $[= 20/(40 - 18a)]$	0.5	0.64	0.91	1.64	5.0	∞	-4.0	-1.43	-0.87 ...

If the lens is very far from the eye, as in the case of a lens mounted in front of a distant instrument dial or TV tube, Eq. (1) tells us that

$$\text{MP} = \frac{mV}{f'(m-1) + E},$$

and if we neglect $f'(m - 1)$ in comparison with E, then M.P. $= mV/E$, approximately. Thus, by moving the object relative to the lens, we change m and M.P. in the same proportion. Note that the lens must be larger than the object we are trying to magnify (Fig. 11.2). The top of

I. THE SIMPLE MAGNIFIER

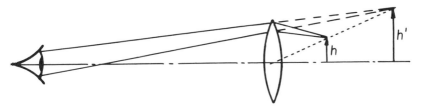

FIG. 11.2. A line through the middle of a thin lens joins object and image.

the object, the top of the image, and the middle of the lens all lie in a straight line.

A. THE BIOCULAR SLIDE VIEWER

The term *biocular* refers to any magnifying system in which the lens is so large that both eyes can see the magnified image simultaneously (Fig. 11.3). This type of viewing system is often used to view color slides. The image is located about 5 in. behind the front of the lens and the magnifying power is about 2.5×. Actually, the magnification generally appears to be greater than this because we are more conscious of area magnification than linear magnification. Thus a linear magnifying power of 2.5 seems to us to be more like $(2.5)^2$, or about 6×. Certainly a color slide magnified in this way is much easier to see than the direct slide, and the details are far more clear.

B. PROJECTION OF AN IMAGE INTO A LARGE FIELD LENS

Projection of an image into a large field lens is often used for greater operator convenience, particularly if a large amount of mate-

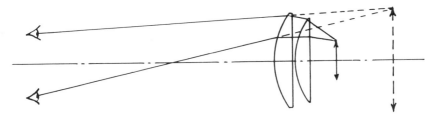

FIG. 11.3. A biocular slide viewer.

rial has to be examined. It permits both eyes to be used, making it a biocular system.

In Fig. 11.4, we see that the object is projected into a large field lens by a high-aperture projection lens of focal length f and relative aperture N. The distance of the image from the projection lens is $L = f(1 + m)$. If the eyes are located at a distance E from the field lens, then the diameter of the exit pupil, which is an image of the projection lens aperture formed by the field lens, is given by

$$EP' = \frac{f}{N}\frac{E}{L} = \frac{E}{N(1 + m)}.$$

The magnifying power of this system is obviously equal to $m(V/E)$.

As an example, suppose we desire an exit pupil with a 2½-in. diameter to accommodate both eyes, and $N = 1.5$ (a practical limit for the relative aperture of a projection lens); then $E = 15$ in., and we find that $m = 3 \times$. Any attempt to increase the image magnification will result in a smaller exit pupil. The overall magnifying power of this system is only 2×, so the system is comparable with the biocular magnifier considered in the last section.

For monocular vision, however, an exit pupil of 1-in. diameter is satisfactory, and in that case the magnification becomes 9 and the magnifying power becomes 6×.

In some types of microfilm or microfiche viewers a concave mirror is used in place of the field lens. The film is at the focus of the mirror, and parallel light enters the observer's eyes.

It should be noted that we can use a large field lens to view a pair of stereoscopic images by using two projection lenses to form two exit pupils side by side, each pupil being about 1 in. in diameter and at a

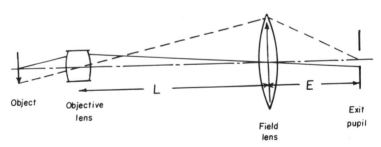

FIG. 11.4. Projection of an image into a large field lens.

FIG. 11.5. Stereoscopic viewing in a large field lens.

separation of about 2½ in. A plan view of the system is shown in Fig. 11.5. The right- and left-hand images are superposed in the large field lens, but each eye sees only its own particular image.

II. THE COMPOUND MICROSCOPE

A. Magnifying Power

If a magnifying power higher than about 10× is required, it is necessary to use a two-lens system, generally referred to as a compound microscope (Fig. 11.6). In this system, the first lens, called the *objective*, forms a magnified aerial image of the object that is then magnified still further by an *eyepiece* situated in front of the eye. The overall magnifying power of the system is given by

M.P. = objective magnification × eyepiece power.

The distance T from the rear focal point of the objective to the front focal point of the eyepiece is called the *optical tube length* of the system. It is generally 160 mm, (but this is not a rigid requirement), and in that case, the objective magnification is equal to $160/f_o$, where f_o is the focal length of the objective lens. The magnifying power of the eyepiece is, given by V/f_{EP}, where V is the conventional nearest

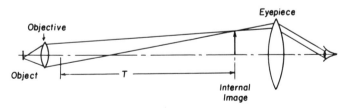

FIG. 11.6. The compound microscope.

distance of distinct vision and f_{EP} the focal length of the eyepiece. Hence, the overall magnifying power of the complete microscope is given by

$$MP = (160/f_o)(250/f_{EP}).$$

For example, if the focal length of the objective lens is 16 mm and that of the eyepiece is 25 mm, each lens contributes 10× to the overall power, which therefore amounts to 100×.

B. Numerical Aperture

It was shown in Section III of Chapter 2 that the resolving power of a perfect lens for incoherent (i.e., self-luminous) objects is given by

$$LRS = \tfrac{1}{2}\lambda/n \sin U,$$

where LRS is the least resolvable separation between close objects on the stage of the microscope, λ the wavelength of the light, angle U the slope of the steepest ray that can enter the objective lens from the axial point of the object, and n the refractive index of the space in which the angle U is measured. The product $n \sin U$ is called the *numerical aperture* of the objective, commonly abbreviated to NA. It is invariant across plane refracting surfaces, so that the NA of the objective is the same whether measured in the air, in the cover-glass over the specimen, or in the Canada balsam in which the specimen is mounted.

To achieve the highest possible resolving power, the wavelength of the light must be reduced and the numerical aperture increased. The largest practical value of the angle U is about 65°, and if this is measured in air, as it generally is, the maximum value of the numerical aperture becomes 0.90, the sine of 65°. Actually, very few microscope objectives reach this value, and a more common limit is about 58°, the sine of which is 0.85.

Most microscopical objects are mounted in Canada balsam under a thin cover glass; in this case the refractive index at the specimen is about 1.51 and the angle U at the specimen drops to 34° because of the refraction of the ray at the uper surface of the cover-glass. Thus, at the object, the numerical aperture is

$$1.51 \sin 34° \simeq 0.85,$$

the same as it is in air.

II. THE COMPOUND MICROSCOPE

In about 1880 it occurred to Ernst Abbe that if the maximum ray slope of 65° could be achieved in the balsam instead of the air, the numerical aperture of the objective could be raised by a factor equal to n, i.e., to 1.3 instead of 0.85. To achieve this, he had to eliminate the air layer between the cover glass and the front lens of the objective to prevent total internal reflection from occurring at the upper surface of the cover glass. This he did by filling the space with oil having a refractive index about equal to 1.5, thus virtually embedding the object inside the front lens of the objective, leading to the term *homogeneous immersion* for this system of operation. Of course, the objective lens had to be specially designed to operate in this unusual fashion; it would be useless to place a layer of oil under any ordinary microscope objective. An oil-immersion objective is shown alongside an ordinary objective in Fig. 11.7.

The other factor involved in the formula for the limiting resolving power of a microscope is the wavelength of the light. If this could be reduced from the normal visible wavelength of about 0.55 μm to an ultraviolet wavelength of, say, 0.28 μm, a further increase in resolution of almost two times could be obtained. This possibility was explored during the early 1900s, but the experimental difficulties were very great. Ordinary glass is opaque to the ultraviolet, so all the optical parts had to be made of quartz. A brilliant source of radiation of the desired wavelength had to be available, and focusing was difficult because the eye is insensitive to those UV wavelengths. Instead, a fluorescent screen was used for the rough focusing, after which a series of photo-

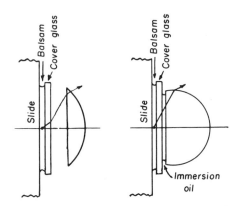

FIG. 11.7. Dry and immersion microscope objectives.

graphs was taken in steps of focus differing by 1 μm, hoping that one of them would be sharp enough for use. Of course, with the coming of the electron microscope in the 1930s, all activity in the ultraviolet ceased.

C. Diameter of the Exit Pupil

From Fig. 11.8, it is clear that the radius r of the exit pupil of a microscope is given by

$$r = f_{EP} \sin U' = \frac{(V)}{m_{EP}} \frac{(NA)}{m_o} = \frac{(V)(NA)}{MP}$$

where MP is the magnifying power of the whole microscope and V the nearest distance of distinct vision, or 250 mm. Thus the diameter of the exit pupil of a microscope is equal to 500 NA/MP. For example, in a 100× microscope equipped with an objective having a numerical aperture of 0.25, the diameter of the exit pupil is 1.25 mm.

Microscopes typically have a very small exit pupil. This has the effect of stopping down the eye so that the spherical aberration is negligible and the resolving power is close to the diffraction limit (see Fig. 10.3). In this case, the eye and the microscope work together to form a single diffraction-limited system.

D. The Vertical Illuminator

If the surface of an opaque object is to be examined under a microscope, some means must be found to illuminate it. At very low powers it is merely necessary to focus a light source onto the object by means of a condensing lens held to one side of the microscope, but at higher powers there is insufficient space for this. In such a case, light must be transmitted down through the objective lens, acting as a

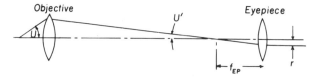

Fig. 11.8. Exit pupil radius of a microscope.

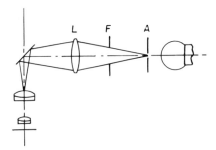

FIG. 11.9. A vertical illuminator.

condenser for bright-field illumination, or through a ring-lens system surrounding the objective for a dark field.

The usual procedure for bright-field illumination is to insert a *vertical illuminator* between the objective lens and the end of the body tube. A possible system is shown diagrammatically in Fig. 11.9. Here the lens L serves two functions—it forms an image of the aperture iris A on the objective lens aperture, in the manner of the first condenser in Köhler illumination, and after passing through the objective lens, it forms an image of the field iris F on the stage. The 45° glass plate shown in the diagram serves as a beam splitter, reflecting perhaps half

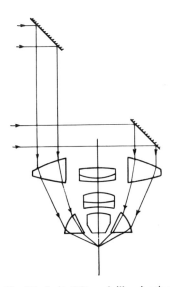

FIG. 11.10. The Leitz Ultrapak illuminating system.

the light and transmitting the other half. At best, therefore, only 25% of the incident light reaches the image, and an intense light source is required, particularly if the object under examination is not very reflective. Microscopes of this kind are used extensively in metallurgy, and also in the microcircuit industry to inspect integrated circuits for gaps and bridges that would make the circuit inoperative.

If oblique illumination is required, the beam splitter in the vertical illuminator can be replaced by a semicircular opaque mirror; this sends light down through one side of the objective, with the reflected light coming up through the other side. For a dark field we may use the Leitz Ultrapak system shown in Fig. 11.10, in which an annular condenser system is built around the objective, light being reflected down through the condenser by means of an annular mirror.

E. Catoptric Objectives

For work in the IR and UV regions, an all-mirror objective is often preferred. It exhibits no selective absorption and no chromatic aberration, so that it can be focused in the visible and used in any wavelength without change in focus. The type suggested by Schwarzschild is described in Section III of Chapter 14.

F. Zoom Microscopes

Recently, many microscopes have been provided with continuously variable magnification by the insertion of a zoom Bravais arrangement between the objective and the eyepiece. This often consists of three lenses, plus–minus–plus, where the first lens is fixed at a convenient distance above the objective lens and the other two lenses are movable in such a way as to change the magnification without shifting the image from the normal eyepiece focal plane. Because the aperture is small (about $f/20$ for a $10\times$, NA = 0.25 objective) and the field is also small (say, $\pm 3°$), fairly simple cemented doublets are often good enough for the lenses in such a zoom system.

A possible system in thin-lens layout is shown in Fig. 11.11, the magnification ranging from unity to $2\times$, as shown. At unit magnifica-

II. THE COMPOUND MICROSCOPE

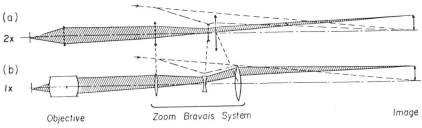

FIG. 11.11. Arrangement of a zoom microscope.

tion, the ray emerging from the zoom system must continue along its incident path with no change in slope angle. As an example, we will assume that the focal lengths of the three lenses are 60, −10.5, and 30 mm, respectively, and that the first fixed lens is at 130 mm from the eyepiece focal plane. If the diameter of the final image is 15 mm and if the objective lens is a 16-mm $f/2$ lens with a numerical aperture of 0.25, then the oblique beam is as shown in Fig. 11.11b. The marginal ray is shown at an exaggerated distance from the axis by dashed lines; the shaded area represents the true oblique beam to scale. Figure 11.11a shows the situation when the zoom system is set at 2× magnification. The dashed lines show that the emerging ray from the zoom now has half the slope angle at which it entered the zoom, since $m = u_1/u_k'$. The positions of the two lenses at a series of magnifications are indicated in Fig. 11.12, which shows that the negative lens moves to the right while the rear positive lens moves to the left. To make a linear scale of magnification, two cams must be used, but the rear lens can be

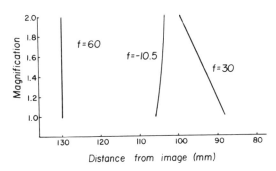

FIG. 11.12. Lens movements in a zoom microscope.

194   11. MAGNIFYING INSTRUMENTS

moved linearly if preferred, in which case only one cam is required. In practice, the user adjusts the magnification by inspection rather than by reading from a scale, and he is not concerned with whether the scale is uniformly divided.

G. Photomicrography[1]

Photographing small objects under a microscope is often quite difficult. Uniformity of illumination over the field is, of course, necessary, and a flat-field objective lens must be used. The eyepiece of the microscope can be merely omitted if a small picture is desired, as on 35-mm film, or a projection eyepiece can be used to enlarge the image. This may be the ordinary eyepiece supplied with the microscope, or it may be a special negative lens called a projection eyepiece, which is not really an eyepiece at all. Both forms are illustrated in Fig. 11.13. In either case, it is necessary to focus the projected image with great care on the film, and the focus setting is usually different from that required by eye because the eye looks at a collimated image, whereas the film requires a sharply focused image at a close distance. The negative projection eyepiece is generally preferred as it tends to help flatten the field.

In any photomicrographic procedure, depth of focus is of paramount importance. Stopping down the aperture of the objective lens increases the depth but tends to enlarge the diffraction image and so reduce the resolving power of the system. Some degree of compromise is therefore required.

H. Microphotography[2]

Photomicrography and microphotography are very different procedures, although the names are similar and often confused. *Microphotography* is the process of forming a tiny image of a pattern or message by working backward through a microscope; the object is placed at the eyepiece end of the tube and the photographic film on the

[1] R. P. Loveland, "Photomicrography," Vols. 1 and 2. Wiley, New York, 1970.
[2] G. W. W. Stevens, "Microphotography", 2nd ed. Wiley, New York, 1968.

II. THE COMPOUND MICROSCOPE

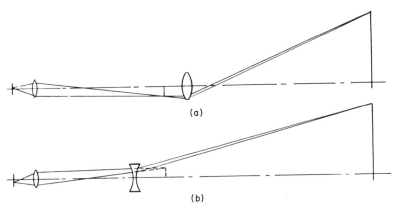

FIG. 11.13. Use of a positive (a) and a negative (b) projection eyepiece for photomicrography.

stage of the microscope, with the illuminant at the eye end. Focusing of the image on the film is the main problem, and sometimes the piece of film is pressed up against a small hole in a metal plate, so that once a good focus has been established, as the result of many trials, other microphotographs can be made without further focusing. Microimages of this kind have been used in the integrated circuit industry and also for secret messages in wartime, where a whole message has been placed on a single period mark in a letter.

I. BINOCULAR MICROSCOPES

Because we generally prefer to use both eyes when observing an object through a microscope, many arrangements for binocular viewing have been devised over the years. These generally involve a beam splitter and prisms forming identical images in the two eyepieces.

For stereoscopic viewing of relatively large objects, two independent erecting microscopes are often mounted together at a convenient angle and focused on the object. Means must be provided for varying the interpupillary separation, and some manufacturers include a zoom power changer in each half.

For high magnifications, it is usual to divide the beam emerging from a single objective lens into two halves by the sharp edge of a prism, where the left side goes to one eye and the right side goes to the other. If the image is erect, the left side of the beam must go to the left eye, but if the image is inverted, the beams must cross over so that the beam from the left side of the objective goes to the right eye, and vice versa. If this is not done, the image will be pseudoscopic.

III. ABBE THEORY OF MICROSCOPE VISION

Abbe realized that most objects seen under a microscope are not self-luminous but must be illuminated by an outside source of light, and the assumption of perfect incoherence that we have made so far is not normally applicable. So he proceeded to investigate what actually happens when a coherently illuminated object is viewed under a microscope.

We consider first the simple case of a ruled grating with alternating black and white bars on a clear background, where the entering light is a plane wave from a collimated monochromatic point source (Fig. 11.14). It is well known from physical optics that such a grating produces a series of spectral orders at slope angles given by

$$\sin \theta = k\lambda/s,$$

where k is a positive integer expressing the particular spectral order, λ the wavelength of the light, and s the grating space, i.e., the distance from one opaque line to the next. The spectral orders form images of the original point source in the upper focal plane of the microscope objective, and light from all these diffraction spectra combines to form the final image seen by the observer.

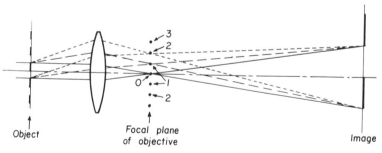

FIG. 11.14. Abbe theory of microscope vision with coherent illumination.

III. ABBE THEORY OF MICROSCOPE VISION

The appearance of the final image depends greatly on the numerical aperture of the objective lens. If the NA is so small that only the undiffracted direct beam, the zero-order, is accepted, then no image will be seen, because the microscope has no way of knowing that there is an object on the stage. If the NA is slightly greater, so that only the first-order spectra are accepted, then the image will be a series of parallel lines corresponding to the lines in the object, but the cross section of intensity will be a sine wave, because a sine-wave grating forms only the first-order spectra and the microscope is obliged to assume that there is such a sine-wave grating on the stage. If the NA is greater, so that more spectral orders are accpeted, then the image progressively approaches becoming a true reproduction of the object, and it is only when all spectral orders are accepted by the objective lens that the image becomes a perfect reproduction of the original object.

Reference to the grating equation shows that, with perfectly coherent illumination, the resolving power of the microscope will be given by

$$\text{LRS} = \lambda/\text{NA},$$

which is exactly half the incoherent resolution given in Section III of Chapter 2.

Abbe performed many experiments to confirm his theory, such as blocking out all the odd-numbered spectra in the upper focal plane of the objective. The line spacing in the image then became half as great as it should be, because a real grating of half the spacing would give the series of spectral orders that were allowed to pass. Today this would be called an experiment in "spatial filtering," but that term had not been introduced in Abbe's time.

In an effort to achieve the same resolution with illuminated objects as with self-luminous ones, Abbe made use of a substage condenser, which formed an image of an extended light source on the stage with an illuminating beam having an NA equal to or greater than the NA of the objective. This situation is indicated in Fig. 11.15. The first-order spectrum of the light entering at one side of the beam is accepted by the objective if its angle of diffraction is twice as large as the case of collimated illumination just considered, thus restoring the resolution to its incoherent value.

This so-called Abbe illumination worked well when the light source was a kerosene flame, which was quite smooth and uniform, but when electric filament lamps were introduced, the filament appeared in

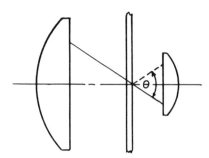

FIG. 11.15. Use of Abbe illumination to double the resolving power of a microscope.

sharp focus in the plane of the object. To eliminate this difficulty, A. Köhler, in 1893, suggested using a two-stage illuminating system similar to that of an ordinary slide projector (Fig. 11.16). A large condenser lens is used to form an image of the lamp filament in the aperture of the substage condenser, which in turn forms an image of the first condenser on the stage. Under these conditions the first condenser would provide a uniformly illuminated disk of light in the plane of the object being examined.

Two iris diaphragms are usually added to the Köhler system, one at the first condenser, to control the size of the disk of light on the stage and thus eliminate stray light, and the other at the substage condenser, to control the numerical aperture of the illuminating beam. This latter diaphragm controls the degree of coherence of the illuminant, with the result that, if the aperture is reduced, the contrast in the image is increased (at the expense of resolution). The reason for this is shown by the two MTF curves in Fig. 2.7.

This is a good place to mention that a microscope is not always used to obtain the ultimate in resolving power. Often a simple increase in

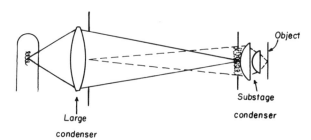

FIG. 11.16. The Köhler illuminating system.

size is all that is required, and if the object is transparent, as biological specimens nearly always are, an increase in contrast is much more useful than an increase in resolution. The practical microscopist spends much time adjusting the iris diaphragms in his instrument, changing their diameter and moving them sideways, in an effort to see as much as possible of the object on the stage.

IV. MICROSCOPY OF TRANSPARENT OBJECTS

A. Dark Field Illumination

It is one of the unfortunate facts of microscopy that most biological objects in thin section are transparent. However, they are not like a piece of jelly, but do have some internal structure, and the problem is to make this structure visible. Some specimens are susceptible to differential staining by suitable dyes; in other cases ultraviolet light can be used to detect structures that absorb some wavelengths more than others. One of the more successful procedures is to use dark-field illumination, in which the object is made to scatter light into the microscope objective while none of the direct light can enter it.

If the numerical aperture of the objective is small, all that is necessary is to place an opaque stop in the middle of the substage condenser aperture, making sure that the stop is large enough to cut out all the direct light that might enter the objective. For objectives of higher aperture a special condenser is necessary, two forms of which are available, the paraboloid and the cardioid (Fig. 11.17). These condensers emit a hollow beam of light, the inside cone angle of which is

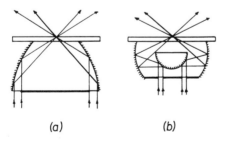

Fig. 11.17. Two types of dark-field condenser.

greater than the acceptable cone of light at the objective. The cardioid form has a wider cone than the paraboloid.

B. The Interference Microscope

If the object under observation is a phase object, i.e., one in which variations in thickness or refractive index cause the transmitted light wave to be a wrinkled instead of a smooth surface, then interfering this transmitted wave with a fraction of the incident wave causes variations in phase to be converted into variations in brightness, so the object becomes visible.

Many forms of interference microscopes have been devised, a typical one being the Dyson microscope, illustrated in Fig. 11.18. This is really a tiny Mach–Zehnder interferometer, with the object situated in one beam. Entering light is split into two beams at the transparent mirror BB', one beam going through the object and the other passing to one side of it. When the two beams are recombined at another beam splitter CC', they interfere, and variations in phase are converted into variations in brightness, as required. An image of the combined waves is formed by a large concave spherical mirror at the top, and this image is examined by an ordinary microscope.

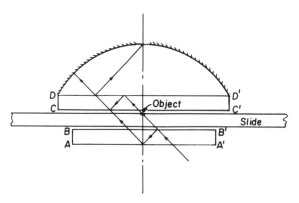

Fig. 11.18. The Dyson interference microscope. [From J. Dyson, *Proc. Roy. Soc. London* **A204**, 170 (1950).]

C. Phase Microscopy

An alternative procedure is the *phase microscope* suggested by Frits Zernike in 1935. This method uses an ordinary microscope with a few special additions. In the lower focal plane of the substage condenser an annular opening with an opaque center is mounted (Fig. 11.19), and a matching annulus is mounted in the upper focal plane of the objective lens; this annulus consists of a ring of quarter-wave retarding material that exactly fits the image of the opening in the lower annulus. Thus all the direct light is retarded in phase by a quarter-wave, whereas the scattered light from the fine structure of the object is unaffected. In the final image these two beams interfere, and a complete analysis shows that variations in phase at the object become variations in intensity at the image, as required.

D. Image Modification

Another method for making a transparent phase object visible is by the use of coherent light and a so-called dc block. Referring back to the beginning of this section dealing with the Abbe theory of microscope vision, it is seen that we can insert a small central obstruction

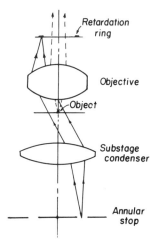

FIG. 11.19. The phase contrast microscope.

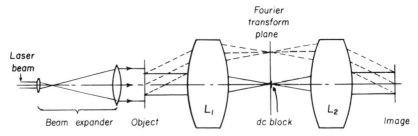

Fig. 11.20. Image processing in coherent light.

into the upper focal plane of the microscope objective, which will cut off all the direct light and yield a dark field when there is no object on the stage of the microscope. Diffracted light from a transparent object passes this obstruction and goes on to form an image, thus converting the phases in the object into variations of intensity in the image plane.

On a large scale, this procedure is known as *image processing,* and the apparatus used is shown in Fig. 11.20. Collimated light from a laser falls on the object which is at the anterior focus of lens L_1. At the posterior focus of this lens is the Fourier transform plane, where the diffraction images are displayed and the small central dc block is located. A second identical lens L_2 then reimages the object at its rear focal plane, but only the diffracted light is used because the direct light has been eliminated. The three planes in this system are located at the focal points of the two lenses; the combined system is afocal and works at a -1 magnification. This is, of course, an example of spatial filtering, but a very simple one. Some much more elaborate forms of filtering are sometimes used in the Fourier transform plane for special purposes such as pattern recognition.[3]

---

[3] Much additional information on the construction of the microscope and its accessories can be found in B. O. Payne, "Microscope Design and Construction." Cooke, Troughton and Simms, York, England, 1957.

CHAPTER 12

The Telescope

I. FUNDAMENTAL PROPERTIES

A. Magnifying Power

A telescope may be regarded as an afocal "black box" that takes an entering parallel beam and converts it into an emerging parallel beam having a different diameter. When such an instrument is used in front of the eye, it changes the angular subtense of a distant object, making the image appear larger or smaller than the original object seen directly.

The change in angular subtense is equal to the ratio of the diameter of the entering beam to the diameter of the emerging beam. The reason for this is indicated in Fig. 12.1. Suppose the plane wave AB from the top of a distant object emerges as the plane wave $A'B'$ at the other end of the instrument. A second plane wave CB originating at the bottom of the distant object emerges as $C'B'$. Because the lag of C behind A is maintained in the passage through the instrument, we see that $CA = C'A'$. The angular subtense of the object is α and the angular subtense of the image is β; hence the angular magnifying power (MP) of the telescope is β/α paraxially, or $\tan \beta / \tan \alpha$ if the angles are large. Evidently, since $CA = C'A'$, the MP is equal to the ratio of D to d, i.e., the compression ratio of the beam on its passage through the instrument. This result is, of course, strictly in accordance with the Lagrange

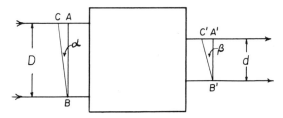

FIG. 12.1. The telescope as a "black box."

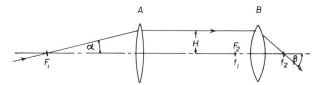

FIG. 12.2. The simple telescope.

theorem, which states that the product of aperture and field must be constant.

The simplest form of telescope consists of two positive lenses separated by the sum of their focal lengths, so there is a common focal point somewhere between them (Fig. 12.2). An oblique ray entering through F_1 at a slope angle α runs parallel to the axis between the lenses and emerges through f_2 at angle β. The ratio of tan β to tan α is thus equal to f'_A/f'_B, which gives the MP, negative in this case because the image is inverted. The first lens A is called the *objective* because it faces the object, and the second lens B is called the *eyepiece* because it is close to the eye.

If we draw an oblique beam of light filling the objective lens, we obtain the situation shown in Fig. 12.3. The internal image lies at the common focal point F of the two lenses. The principal ray of the oblique beam passes through the middle of the objective lens, which therefore serves as both the aperture stop and the entrance pupil of the system. The exit pupil is at EP', where the principal ray crosses the axis after passing through the system; EP' is, of course, a real image of the objective lens formed by the eyepiece, and this is where we must place our eye if we wish to see the whole image. For this reason EP' is sometimes called the *eyepoint*. The size of the field of view is limited by the diameter of the eyepiece, and often a metal ring is placed at the internal real image to form a sharp edge to the field; this ring then becomes the *field stop* of the system.

Since the ratio of the diameter of the entrance pupil to that of the exit pupil is equal to the MP of the telescope, this provides a convenient way to determine the MP of a large astronomical telescope. In such a case the diameter of the objective lens is, of course, known, and the diameter of the exit pupil can be conveniently measured by a little glass scale set at the focus of a magnifying lens, the instrument being known as a *dynameter,* or power measurer. Thus, for example, if the diameter of the objective lens is 30 in. and the measured diameter of the exit pupil 0.25 in., the MP of the telescope is 120×.

I. FUNDAMENTAL PROPERTIES

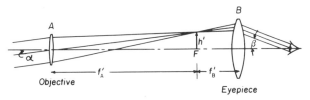

FIG. 12.3. Oblique beam through a simple telescope.

B. Real and Apparent Fields

The real angular semifield of a telescope is the angle α, and the apparent angular semifield is β, both angles expressed in degrees. If the oblique magnifying power $\tan \beta / \tan \alpha$ differs from the paraxial magnifying power f'_A/f'_B, this indicates the presence of some degree of distortion. In practice, most eyepieces have some pincushion distortion that may reach as much as 10% at very wide angles, or 5 to 6% at moderate fields. This is just about the amount of distortion needed to make the actually observed oblique MP equal to the angle ratio β/α rather than the tangent ratio $\tan \beta / \tan \alpha$ that would be theoretically correct in the absence of distortion. Thus, a 6× telescope may have $\alpha = 4°$ and $\beta = 24°$ (and so $\beta/\alpha = 24/4 = 6$), whereas the tangent ratio of 24° to 4° is 6.38, indicating a distortion of $0.38/6 = 0.063\,(100) = 6.3\%$.

In prism binoculars it is customary to express the true field as so many feet at 1000 yards (yd), so a real semifield of 4° covers 420 ft at 1000 yd. Using the same eyepiece and a longer focus objective would give a 7× binocular with a real semifield of 3.43°, or 360 ft at 1000 yd. Some wide-angle binoculars have an apparent semifield of 32°, so if the magnifying power were 7×, the true semifield would be about 4.6°, or 480 ft at 1000 yd.

C. Image Magnification

It was shown in Section III of Chapter 5 that the image magnification of an afocal system is equal to the ratio of the diameters of the emerging and entering parallel beams. This is, of course, the reciprocal of the telescopic MP, and now a paradox is immediately presented: why does a telescope appear to magnify when the actual image is smaller than the object? We can resolve this paradox by recalling that if the object is at a distance L from the anterior focus of the objective lens, the distance of the image from the posterior focus of the eyepiece lens is Lm^2, where m is the image magnification. Thus, in angles, the subtense of a distant object is H/L and the subtense of the image at the eye is H'/Lm^2. But $H' = mH$; hence the image subtense is H/Lm, and the image subtense at the eye is greater than the subtense of the object by the ratio $1/m$, which is the magnifying power of the telescope. Thus, in a 6× telescope, if the object distance is 360 ft, the image distance will be 10 ft. The image height is one-sixth the object height; thus the image of a 6-ft-high object will have a height of 1 ft, and the angular subtense of the image will be six times as great as the angular subtense of the object.

It is interesting to note that a prism binocular can be used back-to-front as a magnifier of small objects. If the object is held at the exit pupil of the instrument, its image will lie in the plane of the objective lens at a magnification equal to the MP of the binocular, where it can be seen by looking into the front end. If the small object is moved toward or away from the eye along the axis of the system, its image will move rapidly along the axis and appear to grow or shrink even though its actual size remains constant; only the distance of the image from the eye is changing.

D. Resolving Power

The resolving power of a telescope can be derived from Rayleigh's rule that two equally bright stars can be resolved if the image of one star falls on the first dark ring in the Airy disk of the other star, i.e., if the separation of the two star images is $1.22\lambda(f/D)$. This relation is derived in Section III of Chapter 2.

However, as explained there, the 1.22 factor can generally be omitted, and in any case, with a telescope we are more concerned with the angular separation of the stars than with the linear separation of their images. Hence, the angular resolution of a telescope becomes simply λ/D rad.

This situation is illustrated in Fig. 12.4, where the plane waves from two just-resolved stars are shown entering the objective of a telescope. The plane waves meet at the bottom and are separated by the wavelength of the light at the top.

In inch measure, the wavelength is approximately 1/50,000 in., and

I. FUNDAMENTAL PROPERTIES

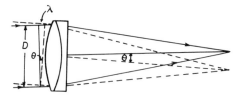

FIG. 12.4. Two stars at the limit of resolution.

if the aperture diameter D is in inches, the resolving power becomes

$$1/(50{,}000\ D)\ \text{rad} = 4/D\ \text{arc-sec}$$

because there are 206,000 sec in a radian. This is known as Dawes' rule and is borne out by small telescopes. However, because of atmospheric turbulence, it is rare that an astronomical telescope can resolve better than about 1 arc-sec, and the larger the telescope the more trouble it has with atmospheric turbulence. It has been said that an aperture of about 8 in. is best from the point of view of resolution.

When a large telescope is placed in orbit in space, we may hope to achieve resolutions approaching Dawes' rule, but the mirror must be made with an accuracy far exceeding the ordinary quarter-wave limit. The mirror will no longer be subject to gravitational distortion and atmospheric absorption in the ultraviolet will not be a problem. The space vehicle must be locked onto the stars with an accuracy comparable to the desired angular resolution to prevent blurring of the image during a long exposure.

E. Brightness of Telescopic Images

We saw in Section II of Chapter 7 that the luminance of an extended object is the same whether we look at it directly or through an optical instrument ($B' = tB$). On the other hand, the image of a star is a point under all conditions of magnification, and starlight therefore piles up on a single retinal detector element in proportion to the area of the telescope objective. Thus if we view a star through a 4-in. diameter telescope, the star will appear brighter in proportion to $4^2/0.25^2$, assuming that the pupil diameter is 0.25 in. under night viewing conditions. Thus, the star will appear to be 250 times as bright when seen through the telescope as compared with direct naked-eye vision.

12. THE TELESCOPE

The sky background, of course, will appear somewhat dimmer through the telescope, which explains why so many faint stars become visible when viewed through a telescope.

Photographically, the situation is similar, for once more the light in a star image is increased in proportion to the area of the telescope objective aperture. The illumination in the sky background now depends on the square of the F-number. Hence, to photograph stars against the sky background, we should have a telescope with a large diameter and also a high F-number, implying a very long focal length.

II. EYEPIECES

An almost endless series of eyepiece designs exists, but by far the largest number of practical eyepieces fall into five classes: the Huygens, Ramsden, Kellner, Delaborne, and Erfle types. The structures of these are shown in Fig. 12.5.

The earliest eyepiece was that proposed by Huygens in 1703. It consisted of two plano-convex lenses spaced apart with their plane surfaces toward the eye, the separation being chosen so as to correct the lateral color, (chromatic difference of MP.) A simple lens used as an

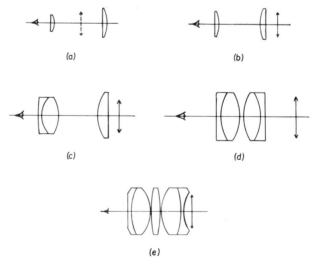

FIG. 12.5. Five typical eyepieces: (a) Huygens (15°); (b) Ramsden (15°); (c) Kellner (20°); (d) Delaborne (25°); (e) Erfle (32°).

II. EYEPIECES

eyepiece suffers from a large amount of lateral color because the focal length is greater for red light than for blue light, and the MP is equal to 10 in. divided by the focal length (see Section I of Chapter 2). Huygens showed that if the separation between two positive lenses is equal to the average of their focal lengths, the lateral color will be corrected. His logic is as follows: the focal length of a separated pair of thin lenses made from the same kind of glass is given by

$$1/F = 1/f_A + 1/f_B - d/f_A f_B$$
$$= (n-1)c_A + (n-1)c_B - dc_A c_B (n-1)^2. \quad (1)$$

To eliminate lateral color, the power of the combination must be stationary as the wavelength is varied, or $\partial(1/f)/\partial n$ must be zero. Differentiating Eq. (12.1) gives

$$\partial(1/f)/\partial n = c_A + c_B - 2dc_A c_B (n-1),$$

and if this equated to zero and multiplied through by $n - 1$, we obtain

$$d = (f_A + f_B)/2,$$

which proves the theorem.

In this analysis no restriction is placed on the focal lengths of the two individual lenses, but for the correction of coma it is desirable to make the field lens (the lens closer to the objective) weaker than the eye lens. A focal-length ratio of 1.5 to 2.3 is suitable for a long telescope, but a smaller ratio, say, 1.4 to 1.8, is better for a microscope eyepiece. Today, Huygens eyepieces are still commonly used for a microscope where the exit pupil is very small, but some kind of achromatic eyepiece is generally preferred for a telescope where the exit pupil is much larger.

In laboratory or surveying instruments where a reticle or cross-wire is required, the Huygens eyepiece is inconvenient because the focal plane is virtual, i.e., inside the system. For this reason Ramsden modified the Huygens eyepiece by making both lenses of the same power and turning the field lens around so the curved sides of the two lenses are adjacent. By reducing the airspace he could move the focal plane outside the system, at the expense of a small amount of lateral color, but this is not generally significant at narrow angular fields.

Prism binoculars have traditionally been equipped with Kellner eyepieces, which are really achromatized Ramsden eyepieces, the eye lens being composed of a cemented doublet instead of a single lens. The angular field of the Kellner eyepiece is about $\pm 20°$ instead of the $\pm 15°$ typical of the Huygens and Ramsden types.

Military optical instruments are often equipped with eyepieces of the Delaborne type, consisting of two thick, strong, cemented doublets with plano external surfaces and curved sides adjacent. This type covers a semifield of 24° with excellent definition.

Finally, the most common wide-angle eyepiece is that patented by Erfle in 1918. It is basically of the Delaborne type, with a strong biconvex element placed between the two cemented doublets. In this way a field of $\pm 32°$ or even more can be covered with acceptable definition. It is used in most present-day wide-angle binoculars.

When designing any instrument embodying an eyepiece, it is essential to provide an adequate *eye relief,* or distance from the back of the instrument to the exit pupil where the eye would normally be located. This distance should never be less than 15 mm, and 20 mm is preferable. If the angular field is very wide, it may be necessary to roll the eye to view the whole field, and in that case it is advisable to increase the eye relief still further, because the exit pupil of the instrument must then be close to the center of rotation of the eyeball rather than at the eye pupil itself.

Another common problem in eyepieces is spherical aberration of the exit pupil. When this is present, the exit pupil for the edge of the field is closer to the eyepiece than the exit pupil for the middle of the field, hence the eye must be brought closer to the eyepiece when the edges of the field are being observed.

III. ERECTING TELESCOPES

In a simple telescope containing only an objective and an eyepiece, the image seen by the eye is inverted, and for terrestrial use some means must be found to erect the image. If the telescope is to be kept short, a prism erector can be used, as discussed in Section VI of Chapter 9. On the other hand, we sometimes require a long telescope, particularly in some military applications, and then a lens erector is much preferred.

A. THE FOUR-LENS TERRESTRIAL EYEPIECE

The oldest means for image erection is the well-known four-lens terrestrial eyepiece, illustrated in Fig. 12.6. The first two lenses constitute the erector, and the last two lenses are an ordinary Huygens

III. ERECTING TELESCOPES

FIG. 12.6. The four-lens terrestrial eyepiece.

eyepiece. By sliding the erector along the axis and moving the eyepiece as needed to keep the image in focus, a variable power telescope can be constructed. Such telescopes were at one time known as pancratic, but today they would be called zooms.

B. THE SUBMARINE PERISCOPE

An outstanding example of a long lens-erecting telescope is the submarine periscope, illustrated in Fig. 12.7. (In this diagram the tube has been divided into three parts for convenience.) Because the top of the tube must be kept narrow with a rather wide angular field, three erectors are used, and there are right-angle prisms at top and bottom to

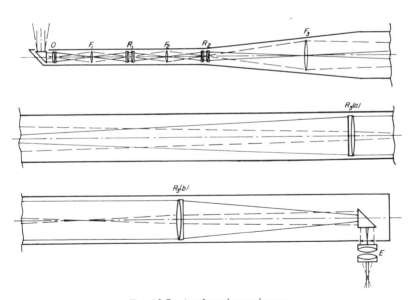

FIG. 12.7. A submarine periscope.

give a horizontal view through a vertical tube. At the top of the tube, after the prism, is an objective lens O covering a field of perhaps ±4° and forming a real image of distant objects at F_1. This image is relayed by a small unit-magnification erector R_1 at F_2 and again by another erector R_2 to form an enlarged image at F_3. After this comes the very long main erector consisting of two widely separated components $R_3(a)$ and $R_3(b)$, as shown. The final image is formed in the focal plane of the eyepiece E. With an overall magnifying power of six times, the apparent field would be about ±24°.

In some examples there is only one upper erector, and there are two long erectors situated in the wider part of the tube. The principle is, however, the same.

In some periscopes the magnifying power can be reduced to 1.5× and the true field increased to ±16° by the insertion of a small reversed Galilean telescope above the top objective lens. The entrance pupil of this added system is located below the top negative component, as indicated in Fig. 12.8.

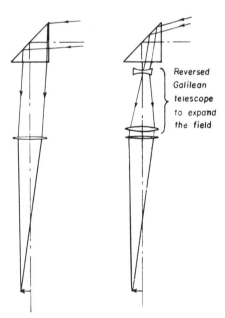

FIG. 12.8. The top of a submarine periscope, with and without a field expander.

C. The Field Lens

It is clear that an oblique beam passing through the objective lens of a periscope might well miss the following relay lens completely. To prevent this, a so-called *field lens* is mounted in the plane of each image, with enough power to form an image of the previous lens aperture on the following lens aperture. Thus the principal ray of an oblique beam crosses the axis in the middle of the objective lens and in the middle of each succeeding erector lens, through the midpoint of the main erector, and finally through the exit pupil where the observer's eye is located.

The presence of a field lens in the plane of an image has no optical effect other than to increase the curvature of field of the final image. The more field lenses that are introduced, the worse the field curvature will become. Indeed, in 1872 it was suggested by C. Piazzi Smyth that a negative field lens could be inserted close to the image plane in a camera to reduce the field curvature of an old-fashioned camera lens. A negative field lens obviously cannot be used in a telescope because it would make the vignetting worse instead of reducing it.

D. A Long Unit-Power Telescope

Another interesting terrestrial telescope is the long unit-power system used by the military to see through a small hole in armor plating (Fig. 12.9). This instrument consists of two identical wide-field telescopes mounted face to face with parallel light between them. The first eyepiece acts as an objective, forming a real image of distant objects at I_1. The two objective lenses then act as a unit-magnification relay to form a second image at I_2, which is viewed through the second eyepiece. The aperture stop is midway between the two objective lenses, and the entrance and exit pupils are outside the two eyepieces.

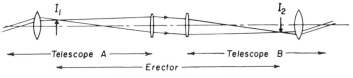

Fig. 12.9. A long unit-power telescope.

Since an exit pupil of 2 or 3 mm in diameter is adequate to fill the observer's eye pupil, the entrance pupil of the telescope also will be only 2 or 3 mm in diameter, because the overall magnifying power of the system is unity. Thus, the instrument can be used behind a small hole in armor plating and still cover a wide field, which may reach 60° or more in diameter if the two telescopes are equipped with wide-angle eyepieces. Unit-power systems of this kind have been made with a length of 6 or 8 ft, both for military applications and to observe inside a closed space in industrial applications.

A similar system has been used in the Gullstrand ophthalmoscope for observing the retina of a patient's eye. The structure of this instrument is indicated in Fig. 12.10. An erect image of the patient's pupil is formed on the doctor's pupil, and an inverted image of the patient's retina is formed on the doctor's retina, which he will see as an upright image. In this way, the doctor can view the whole of the patient's retina at once, or photograph it if need be. Light is provided by imaging an illuminated slit on the patient's pupil via a beam splitter, as shown. To avoid back-reflection which would lower the image contrast, the slit image is usually formed to one side of the portion of the patient's pupil used by the instrument. This is, of course, determined by the size of the erector lens.

E. The Borescope

The borescope is a long narrow telescope intended for use in the examination of the inside of tubes such as gun barrels or the fuel tubes in an atomic reactor. Most borescopes consist of a number of unit-magnification relay units, each perhaps 3 or 4 ft long, which can be

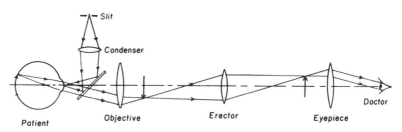

FIG. 12.10. Scheme of the Gullstrand ophthalmoscope.

III. ERECTING TELESCOPES

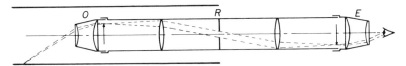

FIG. 12.11. Borescope with telecentric relay.

coupled together to build up the necessary total length. A wide-angle eyepiece O in front forms a real image of the tube wall, which is relayed down to the final eyepiece E, where it can be studied by the observer. To eliminate the necessity for field lenses at the intermediate image planes, the relays are often constructed of unit-magnification afocal units used in a telecentric mode (Fig. 12.11). The length of each unit is four times the focal length of the relay lenses; there is an image at each end and the effective stop is in the middle. Thus the principal ray of an oblique beam enters and leaves each relay unit parallel to the axis and no field lenses are needed.

At the front end, the objective, really an eyepiece with a small entrance pupil, forms a ring-shaped image of a section of the tube wall with a round hole in the center. This unwanted central part of the field is often filled by a plane mirror set at 45° to give a perpendicular view of any part of the tube wall requiring special study. The tube is illuminated by a lamp mounted in front of this 45° mirror. Focusing is accomplished by moving the eyepiece in and out.

F. THE COMMON RIFLE SCOPE

The common rifle scope consists of an objective lens with a reticle or cross-wire at its focus, an erector lens working at about unit magnification, and a large eyepiece. The exit pupil is located as far from the eyepiece as possible to avoid injury due to the recoil of the rifle when fired.

To eliminate vignetting over a moderately wide field, a field lens is introduced at or near the plane of the first image to form an image of the objective lens aperture on the erector lens, so that the principal ray crosses the axis at the middle of the objective (the entrance pupil), the middle of the erector (an internal pupil), and finally at the eyepoint (the exit pupil).

A typical rifle scope is illustrated in Fig. 12.12a, in which the focal

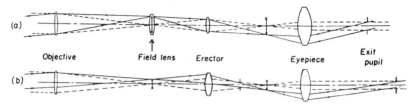

Fig. 12.12. Two types of rifle scope.

length of the objective lens is 5 units, of the erector 1.5 units, and of the eyepiece 2 units. Since the erector is working at unit magnification, the overall magnifying power of the system is 2.5×, and assuming thin lenses, the length from objective to eyepiece is 13 units. As shown, the true field is ±5.2° (tan = 0.09) and the apparent field is ±12.7° (tan = 0.225). The eye relief, or distance from the eyepiece to the exit pupil, is 3.33 units. Pertinent data of the four lenses are listed in Table I. In practice, of course, the thin lenses assumed in the table would be replaced by properly designed achromats, and the erector and eyepiece would probably be made of two cemented doublets placed close together. The field lens could be a single element because it contributes nothing to the aberration correction.

For a rifle scope of minimum cost, it is customary to omit the field lens and greatly reduce the angular field, as shown in Fig. 12.12b. Also, to minimize the vignetting, the erector lens is often made as large as the objective lens. In this example the angular field has been reduced to ±2.9° (tan = 0.05) and the apparent field to ±7.1° (tan = 0.125). At this obliquity the upper rim ray enters at a height of 0.52, which is the aperture of the axial beam, but the lower rim ray enters at a height of only −0.2 in order to strike the erector at the same height of 0.52. We

TABLE I

Data of Lenses in Rifle Scope

Lens	Focal length	Minimum diameter	Relative aperature
Objective	5.0	1.04	$f/4.81$
Field	1.88	0.90	$f/2.08$
Erector	1.5	0.62	$f/2.40$
Eyepiece	2.0	1.92	$f/1.04$

TABLE II

Data of Lenses in Low-Cost Rifle Scope

Lens	Focal length	Minimum diameter	Relative aperature
Objective	5.0	1.04	$f/4.81$
Erector	1.5	1.04	$f/1.44$
Eyepiece	2.0	1.53	$f/1.31$

cannot now refer to a "principal ray" because there is no definite stop; instead we speak of the chief ray of an oblique beam which enters midway between the upper and lower rim rays, namely, at a height of 0.16 units. This ray crosses the axis at a distance of 1.65 beyond the erector, and again at a distance of 4.95 units beyond the eyepiece, where the observer's eye would have to be placed to see the whole field at once. Thus, eliminating the field lens has had the effect of reducing the angular field and moving the exit pupil back from the eyepiece; it has also called for a larger erector and has reduced the oblique beam emerging from the eyepiece to be only about 70% of the height of the axial beam.

Pertinent information on the three lenses as they are now is given in Table II. Since this is an economy design, the three lenses are commonly made from simple cemented doublets, although the apertures of the erector and eyepiece are really too large for such a simple treatment.

G. Telescope with a Stepped Power Changer

A simple form of variable-power telescope makes use of a stepped power changer. There are two positions that a lens can occupy between an object and its image — the A position giving a certain magnification and the B position an identical demagnification. Thus, if A gives a magnification of $\sqrt{2}$ and B a magnification of $1/\sqrt{2}$, the change in magnification when the lens is moved from A to B would be $2:1$.

It is possible to use such a stepped power changer as the erector in a telescope, so that the user can change power by merely moving the erector from one position to the other without having to move the

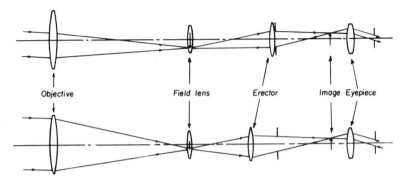

FIG. 12.13. Telescope with a (a) 5× and (b) 10× stepped power changer.

eyepiece. To maintain a constant exit pupil, we can place a stop at the position of the erector in its low-magnification setting, and we must then choose a field lens that images the objective lens aperture into this stop at the high-magnification setting (Fig. 12.13).

As an example, consider a telescope in which the two magnifying powers are 5× and 10×. Assuming that the focal length of the objective lens is 8 cm and the erector magnification is $\sqrt{2}$ and $1/\sqrt{2}$ at the two positions, then the eyepiece focal length must be $8\sqrt{2}/10 = 1.131$ cm. If the erector focal length is 2 cm, its conjugate distances must be $2(1 + \sqrt{2})$ and $2(1 + 1/\sqrt{2})$, respectively.

To determine the power of the field lens, we set up the front end of the high-power system in reverse order, starting at the fixed stop and ending at the objective lens, leaving the power of the field lens to be filled in later (as shown in Table III). The ray shown in this table is a paraxial principal ray traced from the middle of the stop as far as it will

TABLE III

DETERMINATION OF THE FIELD-LENS POWER

	Stop	Erector	Field lens	Objective
$-\phi$	0	-0.5	?	-0.125
d		1.414214	3.414214	8.0
y	0	-0.1414214	-0.2414214	0
u	-0.1	-0.1	-0.0292893	0.0301777

III. ERECTING TELESCOPES

go, namely, to the y of the field lens. As this ray must pass through the middle of the objective lens, the value of u' must be equal to $y/8 = 0.0301777$. Hence, the power of the field lens is given by $(0.0292893 + 0.0301777)/(0.2414214) = 0.2463203$, and the problem is solved.

As a check, we may now set up the whole system, still in reverse order, and trace a principal ray and a marginal ray at each of the two possible positions of the erector. The height of entry of the reverse marginal ray is assumed to be 0.15 cm at either power, and the apparent angular field is ± 0.3. Lastly, at each erector position the heights of the upper and lower rim rays have been found by

$$y_{\text{upper}} = y_{\text{prin}} + y_{\text{mar}} \quad \text{and} \quad y_{\text{lower}} = y_{\text{prin}} - y_{\text{mar}},$$

and the two rays have been added to the diagrams in Fig. 12.13. The ray traces are shown in Table IV. Pertinent information on the four lenses as they are now is given in Table V. The exit pupil diameter is 3 mm and the entrance pupil diameter is therefore 30 mm at the high power and 15 mm at the low power. The angular fields are $\tan^{-1}(0.03) = \pm 1.72°$ at high power and $\tan^{-1}(0.06) = \pm 3.43°$ at low power. The apparent field is thus about $\pm 17°$ at either power. The strongest lens is the erector and it would call for very careful design. The shift of the erector from one position to the other is only $20/\sqrt{2} = 14.14$ mm. The eyepiece diameter is a little larger than its focal length, but this is usual in eyepieces. The objective aperture of $f/2.7$ is excessive, and it is clear that the exit pupil should be made smaller than 3 mm in such a telescope.

Another form of stepped power changer is a low-power afocal Galilean telescope that can be inserted, when required, into a parallel beam of light at any convenient place in a system. The power changer at the top of a submarine periscope shown in Fig. 12.8 is an example of this arrangement. Of course, the Galilean telescope can be inserted into the beam with the positive lens in front, in which case it will have the effect of increasing the MP and reducing the angular field of the instrument. Sometimes the Galilean telescope is mounted into a hollow box which can be indexed around through 180° in order that the power changer can be inserted either way round, or indeed at 90° rotation both lenses can be removed from the system. In this way, if the Galilean telescope has a magnifying power of $\sqrt{2}$, we have three stepped powers available, namely, 0.7, 1.0, and 1.4 times the basic power of the instrument.

TABLE IV[a]

PARAXIAL RAYS THROUGH A TWO-POWER TELESCOPE

	Exit pupil		Eyepiece		Image		Stop		Erector		Field lens		Objective
$-\phi$	0		−0.883883		0		0		−0.5		−0.246320		−0.125
d		1.50627		1.13137		3.41421		1.41421		3.41421		8.0	
						High-power system							
						Principal ray							
y	0		0.4519		0.3394		0		−0.1406		−0.24		0
u		0.3		−0.0994				−0.0994				−0.0291		0.03
						Marginal ray							
y	0.15		0.15		0		−0.4527		−0.6402		0		1.5
u		0		−0.1326				−0.1326				0.1875		0.03
						Upper and lower rim rays							
UR	0.15		0.602		0.339		−0.453		−0.781		−0.24		1.5
LR	−0.15		0.302		0.339		0.453		0.500		−0.24		−1.5
d		1.50627		1.13137		3.41421		0		4.82843		8.0	
						Low-power system							
						Principal ray							
y	0		0.4519		0.3394		0		0		−0.48		−0.3294
u		0.3		−0.0994				−0.0994				−0.0994		0.06
						Marginal ray							
y	0.15		0.15		0		−0.4527		−0.4527		0		0.75
u		0		−0.1326				−0.1326				0.0937		0
						Upper and lower rim rays							
UR	0.15		0.602		0.339		−0.453		−0.453		−0.48		0.421
LR	−0.15		0.302		0.339		0.453		0.453		−0.48		−1.079

[a] All numbers are rounded off to four significant figures.

TABLE V

DATA OF LENSES IN A RIFLE SCOPE

Lens	Focal length (mm)	Clear aperture (mm)	Relative aperture
Objective	80.0	30.0	$f/2.7$
Field	40.6	9.6	$f/4.2$
Erector	20.0	14.1	$f/1.4$
Eyepiece	11.3	12.1	$f/0.93$

IV. OTHER TYPES OF TELESCOPE

A. THE GALILEAN TELESCOPE

In a Galilean telescope a positive objective lens is followed by a negative eyepiece. The path of an oblique beam through the system is shown in Fig. 12.14, giving an erect image. The figure shows that the aperture stop and exit pupil now coincide at the pupil of the user's eye, and the entrance pupil is the virtual image of the eye formed by the instrument at a point behind the observer's head. The field stop is now the objective lens aperture, and the field of view will have a fuzzy edge because the entrance pupil is so far from the field stop. Galilean telescopes have been used extensively as small terrestrial telescopes, opera glasses, telescopic spectacles, and sports glasses, because of their

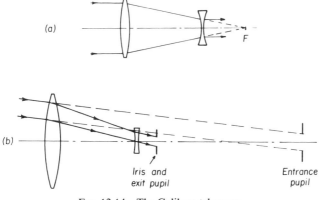

FIG. 12.14. The Galilean telescope.

simplicity, light weight, and erect image. As the angular field is extremely limited, very few Galilean telescopes exceed a 3× MP, although Galileo himself made some telescopes of this type having a magnifying power as high as 30×, with which he observed the moons of Jupiter and the rings around Saturn. The field of view of these telescopes must have been incredibly narrow.

A very small high-powered Galilean telescope made with simple lenses is often mounted in the door of a hotel room to enable the occupant to see who is in the passage outside the room. The strong negative eyepiece lens faces the outside and the positive objective lens is inside the room, so the telescope is used backward and exhibits a greatly reduced image. The telescope is also deliberately defocused, so that on looking into the room the field of view is small and the image is blurred, whereas on looking out through the door the field is very wide and the image is acceptably sharp, although afflicted with a large amount of barrel distortion giving the familiar "fish-eye" effect.

Another common application of a reversed Galilean telescope is in the eye-level viewfinder fitted to practically every camera except the SLR. The structure of this viewfinder is indicated in Fig. 12.15. The large negative lens faces the scene and forms a diminished virtual image of it, as shown by the dashed line at the left of the figure. This image is located at the common focus of both lenses, with the small positive lens P near the eye weaker than the negative lens N, giving an overall magnifying power less than 1.0 (a common value is 0.4×). The angle U represents the semifield of the camera at the top of the format, with the chief ray of an oblique beam entering the eye at a slope angle U', as shown. If it is important to eliminate distortion in this type of viewfinder, the rear surface of the negative lens can be made ellipsoidal rather than spherical.

A reversed Galilean telescope has been used often as a laser beam expander. The beam from a laser is very narrow and in many applications a wide beam is required. The strong negative lens of a Galilean

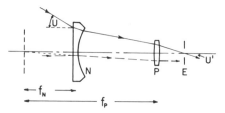

FIG. 12.15. The reversed Galilean viewfinder.

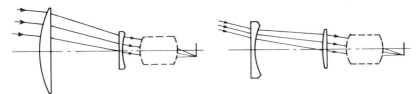

FIG. 12.16. Afocal Galilean attachments for a small movie camera.

telescope expands the laser beam until it is as large as needed, and then the positive lens in the Galilean telescope recollimates the beam. The cross section of a laser beam is not uniform in intensity but has a Gaussian distribution; it can be made more uniform if strongly aspheric lenses are used in the beam expander.

Another application of small Galilean telescopes is as a focal-length changer on a small movie camera. The action of a direct and a reversed Galilean system when mounted in front of a regular camera lens is shown in Fig. 12.16. The direct Galilean attachment increases the focal length of the camera lens by the magnifying power of the telescope, and likewise, the reversed Galilean system shortens the focal length and widens the angular field of the camera lens. Systems of this kind are no longer used because of the general adoption of zoom lenses on small movie cameras. This application is similar to the stepped power changer described in Section III.G.

B. The Donders Telescope

In about 1880 F. C. Donders suggested a three-lens afocal telescope consisting of a pair of positive lenses with a negative lens between them. If the positive lenses have the same power and the separations are equal, the system is symmetrical and the telescopic magnifying power (MP) is unity (Fig. 12.17b). However, on moving the middle lens along the axis the MP increases or decreases. For small movements of the middle lens the system remains substantially afocal, but if the middle lens is moved through a considerable distance it is necessary to refocus the system by moving one of the outer positive lenses in an in-and-out manner, with the maximum length of the system occurring at the symmetrical midposition. For example, if the positive lenses have a focal length of 4 in. and the negative lens a focal length of -1 in., then in the symmetrical case the separations will each be 2 in. If

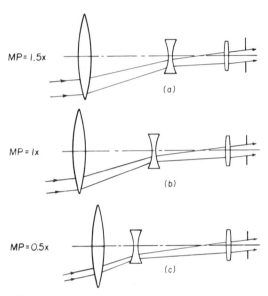

FIG. 12.17. The Donders variable-power telescope.

the middle lens is now moved along the axis by 0.5 in. toward the rear (Fig. 12.17a), the front lens must be moved in by 0.1667 in. to restore the afocal condition. The MP of the system, to an eye situated at the right-hand end, is now 1.5×. On the other hand, if the middle lens is moved by 0.5 in. to the left (Fig. 12.17c), the front lens must be moved in by 0.5 in. and the MP becomes 0.5×. Thus, by these movements, we can construct a zoom telescope having an overall MP range of 3 : 1. As in the Galilean telescope, the aperture stop and the exit pupil are at the observer's eye, and the front lens diameter must be quite large; it constitutes the field stop of the system.

C. Teinoscope

In about 1835 Sir David Brewster suggested making a telescope using only plane refracting surfaces. His argument was based on the "black box" idea outlined in Section I.A, because a telescope is any device that will change the diameter of a parallel beam of light without destroying its parallelism. At minimum deviation, a refracting prism receives and emits a parallel beam without changing its diameter (Fig. 12.18b), but away from minimum the diameter of the beam is indeed

IV. OTHER TYPES OF TELESCOPE 225

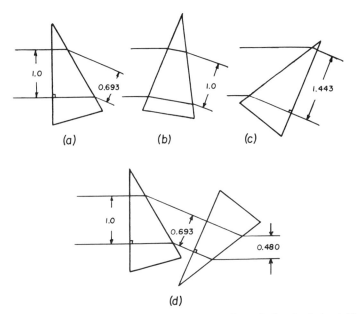

FIG. 12.18. Beam compression caused by a 30° prism of refractive index 1.60.

changed (Figs. 12.18 a and c). Consequently, if we look at a distant scene through a refracting prism and turn the prism out of minimum deviation, we see the scene deviated laterally and highly colored, but also stretched or compressed in a direction perpendicular to the refracting edge of the prism. The apparent size of the scene in the perpendicular direction is unaffected, leading to what is known as *anamorphic distortion.*

However, as Brewster pointed out, both the deviation and the color fringing can be removed by the use of a second prism following the first, one prism being turned clockwise and the other counterclockwise through such angles as to keep the image in its original direction (Fig. 12.18d). An anamorphic system of this kind has been used to decompress the anamorphically compressed image on the film in the CinemaScope system of motion pictures. Prismatic anamorphotes are not suitable for wide angular fields, but they cause little trouble in projection systems. At wide field angles some signs of color fringing appear at the edges of the field; this can be removed by achromatizing the prisms.

Brewster also suggested that the anamorphic distortion could be eliminated by the use of two sets of tilting prisms used in succession, mounted at right angles to one another, so that the image would be stretched or compressed equally in the two perpendicular meridians, making what we would now call a zoom telescope.

D. Endoscopes

Endoscope is a generic term covering a wide variety of optical systems intended to enable a surgeon to observe diseased structures through natural or artificial apertures in the human body. For many years endoscopes were little periscopes made up of a series of relay and field lenses, with a small electric bulb at the far end to provide illumination inside the body, and an eyepiece at the near end. Prior to the introduction of antireflection coatings in the 1930s, the numerous interreflections between lens surfaces caused such a huge amount of stray light that the image was almost unrecognizable. Even with coated lenses the loss of light was still serious, considering how little light was available anyway.

Since about 1958 endoscopes have been constructed with an ordered fiber-optic bundle to conduct the image to the eyepiece, instead of the series of little lenses used previously. A small objective lens and an Amici prism at the lower end form an image on the end of the fiber bundle and the eyepiece is focused on the other end to observe the image. Instead of a tiny electric lamp as a source of illumination, a set of thicker fibers surrounding the imaging bundle is used to conduct light from a large external light source down the tube. This new endoscope structure has led to a vast improvement in image quality and brightness, so it is now even possible to make color movies of internal organs through the instrument.

E. Is It a Telescope or a Microscope?

Doubt sometimes arises as to whether a given instrument is a small telescope or a large microscope. The criterion to be adopted is usually whether or not you have access to the object. If you can get as close as you please to the object, then your magnifying instrument must be regarded as a microscope. On the other hand, if the object is inaccessible and at a fixed distance from the observer, then the instrument must be regarded as a telescope. The meaning of "magnifying power" in the two cases must be carefully defined if it is to be significant. Occasion-

V. ASTRONOMICAL TELESCOPES

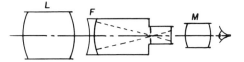

FIG. 12.19. An infrared telescope.

ally a telescope is provided with a sufficiently long focusing range to enable it to be focused on quite close objects; in such a case it may well be called a "tele-microscope."

F. AN INFRARED TELESCOPE

An infrared telescope usually consists of an electronic tube having an infrared-sensitive electron-emitting coating at the front end and a green phosphor at the rear. An electron lens inside the tube forms an image of the emitted electrons on the phosphor at a magnification of -0.5. The front surface of the tube is concentric about the electron lens, and the phosphor is flat, as indicated in Fig. 12.19. A high-aperture ($f/2$ or $f/1.5$) photographic objective L is used to form an inverted image of outside objects on the front face of the tube, generally with a negative field flattener F to produce the required backward-curving image. At the other end of the tube is a magnifier M for viewing the phosphor. (This is often referred to as an eyepiece.) If the objective lens has a focal length of 2 in. and the magnifier a focal length of 1 in., then the overall magnifying power of the device is unity. Both lenses must be focusable, the objective lens to focus on objects at different distances and the magnifier to suit the observer's vision. An infrared telescope is often mounted alongside a source of infrared radiation so that objects can be viewed in the dark by using this instrument.

V. ASTRONOMICAL TELESCOPES[1]

Strictly speaking, an astronomical telescope is not a telescope at all in the sense we have been considering but merely a large image-forming device. The image so formed can be observed through an eyepiece, as

[1] H. C. King, "The History of The Telescope." Sky Publishing, Cambridge, Massachusetts, 1955.

in any small telescope, or it can be photographed (with or without the aid of an image intensifier), or it can be made to fall on the slit of a spectrograph, or on a photometer, or a television camera, or any other suitable receiver.

In all large telescopes the primary image former is a concave mirror, but in smaller telescopes, say up to about 6 inches in diameter, an objective lens is often used. The relative advantages of a mirror and a lens are discussed at the beginning of Chapter 14.

The largest refracting telescopes are the Yerkes (40-in.) and the Lick (36-in.), both made by the Alvan Clarks in the 1890s. With a focal length of, say, 60 ft, the telescope tube is very long, and it must be pivoted about its center for mechanical balance, so the eyepiece or photographic plate holder or spectrograph will rise and fall some 30 ft when going from the horizon to the zenith. This problem is usually overcome by the use of a rising floor that can be set at any convenient height by the operator.

Because the whole sky and everything in it rotates once every 24 hours about a point in the sky which is an extension of the earth's axis of rotation, most astronomical telescopes are mounted on a rotating *polar axis* directed toward the pole of the sky, which is near but not exactly coincident with the star Polaris. The telescope must also have a second axis of rotation perpendicular to the polar axis known as the *declination axis* so that it can be directed toward any desired point in the sky. Such a mounting is called *equatorial.* Recently, a very large and heavy Russian telescope was mounted with horizontal and vertical rotation axes like a theodolite; this is called an *altazimuth* mount. To follow any particular stellar object it is necessary to rotate the telescope about both axes simultaneously, and a digital computer is employed to produce the desired motions in real time.

A. The Coronagraph

The coronagraph was developed by B. Lyot in 1931 in an effort to observe the solar corona without a total eclipse. The problem is that the sun's luminance is about a million times brighter than the corona, and the slightest amount of stray light from the sun will entirely drown the faint light from the corona.

Lyot's achievement was to find ways to eliminate all traces of stray

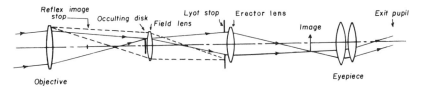

FIG. 12.20. A coronagraph.

light in the instrument. His telescope was constructed on the plan indicated in Fig. 12.20. A black metal occulting disk was placed in the focal plane of the objective to block out the direct image of the sun formed by that lens. The image of the corona surrounding the sun's image would, of course, pass the occulting disk and be relayed to the eyepiece by the erector lens. Unfortunately, there are several sources of stray light besides direct light that had to be eliminated, the largest being diffraction at the rim of the objective lens. Lyot removed this source by placing a field lens immediately behind the occulting disk, to form an image of the rim of the objective lens on an annular shield shaped like a washer, the hole in the shield being a little smaller than the image of the objective-lens aperture. This served effectively to block out diffraction at the rim. The next source of stray light was an image of the sun caused by two reflections within the objective lens itself; this was blocked by a small metal stop placed where this image is formed. The last source of stray light to be considered was scatter at the polished surfaces of the objective lens, and Lyot found that this could be removed by polishing the lens surfaces for several times as long as usual. For the objective lens he used a single element of selected borosilicate crown glass with no trace of bubbles or striae; the aberrations of this simple lens were corrected in the erector after the direct image of the sun had been removed by the occulting disk.

CHAPTER 13

Surveying Instruments

I. CLASSES OF SURVEYING INSTRUMENTS

Large-scale land surveying is generally performed by first establishing a horizontal base line of known length with great care and accuracy. An easily identifiable distant point, such as a hilltop, is next chosen, the angles between the base line and this point measured at each end of the base, and a triangle constructed with the distant point at its apex. The triangle is then projected down to a sea-level plane for map making. The lengths of the sides are calculated, and they in turn serve as the bases of other triangles until the whole area has been covered. The topographic details within each triangle are filled in later, either by local surveying or from aerial photographs. The original surveyed points, called *bench marks,* act as known locations to determine the scale and tilt of aerial photographs. In some cases, surveyors make use of rangefinding devices to determine the distance of a remote object, but these are not always sufficiently precise, except for topographic work.

An independent branch of surveying is concerned with the determination of heights above sea level and the plotting of contour lines. This process is known as *leveling,* and again it is now generally performed by using stereoscopic pairs of aerial photographs.

Classical surveying instruments fall into several classes. The simplest is the *plane table.* Next comes the surveyor's *transit* and the more elaborate and precise *theodolite.* Lastly comes the surveyor's *level,* which is a device for sweeping out a horizontal plane in space. Besides these, we can include the various types of *sextants,* which are hand-held angle-measuring devices.

A. THE PLANE TABLE

The plane table is a horizontal drafting board mounted on a tripod along with a movable alidade consisting of a straightedge on which is mounted a simple pair of sights or small telescope. The measured base line is first marked on the drawing paper to scale, sights are made on a distant object, (such as a church tower), from each end of the base line, and lines are ruled along the straightedge to close the triangle. Other conspicuous objects are then sighted, and more triangles are drawn in. Plane-table work is obviously not very accurate, but it is quick and convenient for small-scale detailed surveys on fairly level ground.

B. The Transit

A transit is a small telescope mounted on a tripod, with vertical and horizontal axes so that the telescope can be sighted on any desired object. A horizontal angle scale is provided so that azimuth angles can be read off, generally to an accuracy of 1 arc-min. Transits are often used for plotting highways and lot lines, where their accuracy is quite adequate.

C. The Theodolite[1]

The name "theodolite" is generally reserved for a much larger and more precise form of transit, equipped with two divided circles to read angles in both azimuth and elevation, often to an accuracy of 1 arc-sec. Theodolites are used for large-scale surveys and for establishing primary triangles over a wide area of land.

D. The Cinetheodolite

This is a special-purpose instrument used to track a fast-moving object such as a rocket. Photographs of the target are made on motion-picture film at a fixed rate, and the altitude and azimuth readings are recorded on each frame. A small telescope is provided so that an observer can keep the instrument sighted on the moving target.

---

[1] H. E. Torberg, W. J. Rowan, and J. R. Vyce, "Optical instruments for metrology," *in* "Handbook of Optics" (W. G. Driscoll and W. Vaughan, eds.), Section 16. McGraw Hill, New York, 1978.

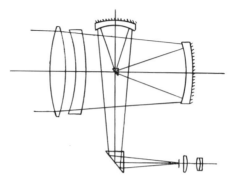

Fig. 13.1. Kern mirror telescope for a theodolite.

E. A Balloon Theodolite

This is another special form of theodolite intended for the observation of sounding balloons or other high-altitude objects. The objective lens is mounted between trunnions, and the light is reflected out through one of the supports to a fixed eyepiece. A particularly compact and complicated instrument is that made by the Swiss company Kern, shown in Fig. 13.1.

Surveying instruments are often equipped with internal focusing devices which are waterproof and more rugged than normal eyepiece focusing arrangements.

F. The Level

The surveyor's level is a small telescope capable of being rotated about a vertical axis in such a way as to sweep out a horizontal plane through the instrument location. For this to be possible the axis of rotation must be precisely vertical and the line of collimation joining the nodal point of the objective lens to the cross-wires in the eyepiece must be exactly perpendicular to the vertical axis. However, as both these conditions are virtually impossible to achieve, it is customary to equip the telescope with a sensitive spirit level running along its length and with a tilting screw beneath the eyepiece so that the operator can set the telescope truly horizontal before making an observation. This

I. CLASSES OF SURVEYING INSTRUMENTS

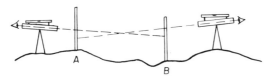

FIG. 13.2. Testing a surveyor's level.

procedure assumes that when the bubble is in the middle of its tube the telescope axis is horizontal, and to check this adjustment the instrument is placed beyond the end of a pair of surveying staffs, the bubble is centered in its tube, and the difference in level between the two staffs is read off. The instrument is then moved to a position beyond the other end of the pair of staffs and the reading is repeated (Fig. 13.2). The two height differences will be identical only if the telescope is truly horizontal when the bubble is centered. If not, the two little adjusting screws that support the bubble above the telescope must be reset, and the experiment repeated until the height difference does read the same no matter where the instrument may be located.

G. The Autoset Level

Recently a new type of surveyor's level has been developed in which the line of sight remains accurately horizontal even if the telescope axis is not precisely level (see Fig. 13.3). The telescope axis joining the objective lens and the cross-wire is shown by a dashed line and is tilted

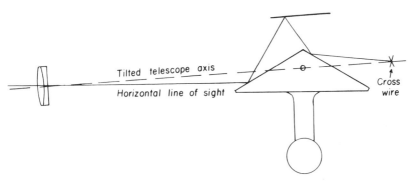

FIG. 13.3. The autoset level.

downward somewhat. There is also a three-mirror Abbe-type invertor which serves to erect the image. The top mirror is attached to the telescope tube, so it also is tilted downward, but the other two mirrors are fastened to a glass block which is pivoted at O and held horizontal by a small weight acting like a pendulum. In this situation, a ray entering horizontally through the middle of the objective lens passes through the cross-wire after emerging from the Abbe erector, just as if the telescope were truly horizontal. There has to be some kind of damping mechanism in the instrument to prevent the pendulum from oscillating during observation, and the pivot O has to be frictionless. (This can be achieved by the use of a flexible metal suspension in place of a regular pivot.) The image is erect but reversed left to right.

H. The Optical Square

A very simple device for establishing two lines perpendicular to each other, such as the two sides of a field, consists of a pentaprism mounted on a tripod. By viewing one of the lines through the prism and the other over it, the two lines will appear coincident if they are truly perpendicular, assuming of course that the pentaprism has been made accurately. The pentaprism is a constant deviation system, so that it does not have to be mounted accurately on its tripod. Such an arrangement is known as an *optical square* by analogy with the familiar carpenter's square.

I. The Sextant

The Hadley sextant is used extensively on shipboard to measure the angular height of the sun above the horizon. It consists of a frame carrying two mirrorss (Fig. 13.4), mirror B filling the whole aperture of a small telescope T and mirror A filling only half the aperture, so the observer sees the sun and the horizon in apparent superposition. By tilting mirror B, the sun's image can be made to rise or fall until it is just bisected by the horizon, at which point the sun's elevation can be read off the arc S. Of course, it is necessary to take into account the refraction of the atmosphere, so the sun's image should just touch the horizon instead of being bisected by it. One clever feature of the sextant is that the sun's image is seen in a constant-deviation combina-

I. CLASSES OF SURVEYING INSTRUMENTS

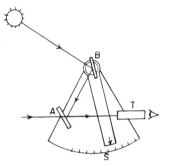

FIG. 13.4. The Hadley sextant.

tion of two mirrors, so the apparent coincidence of sun and horizon is not affected by any random up-and-down movements of the instrument such as might occur on the deck of a ship.

J. THE BUBBLE SEXTANT[2]

In the early days of aviation it was necessary to measure the elevation of the sun rapidly even when no horizon was visible, and instruments called *bubble sextants* were devised in which a small artificial horizon was included within the instrument. This took the form of a small circular spirit level, the bubble and the sun being viewed in coincidence. One possible arrangement is shown schematically in Fig. 13.5. The bubble is formed beneath a spherically curved cover glass B, the center of curvature being at the collimator lens L so that the bubble

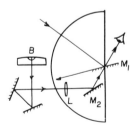

FIG. 13.5. An aircraft bubble sextant.

[2] L. B. Booth, "The aerial sextants designed by the Royal Aircraft Establishment." *Proc. Opt. Conv.*, **2**, 720, (1926).

appears at infinity. The sun is viewed by reflection at the tiltable mirror M_1.

II. RANGEFINDERS

A. STADIAMETRIC

In small-scale suveying it is often convenient to determine the distance of some object directly, and in such a case a stadiametric procedure may be used. Suppose we have a small telescope equipped with a pair of cross-wires in the focal plane which are separated by a distance d. On observing a distant surveyor's staff, the cross-wires are seen to intercept a distance D on the staff. Then obviously

$$R = (D/d)f,$$

where R is the range and f the focal length of the telescope objective. The value of R so obtained is actually the distance from the anterior focus A of the objective lens to the staff, where A is called the *anallatic point*. It is usual to apply a correction of amount $f + x$ to the calculated result, where $f + x$, shown in Fig. 13.6, is known as the *instrument constant*. This is often about 1 ft, and it represents the distance from the anterior focus of the telescope objective to the vertical rotation axis of the instrument. Since the principal ray is parallel to the axis after passing through the instrument, focusing may be accomplished by a movement of the cross-wires and eyepiece together without introducing any error.

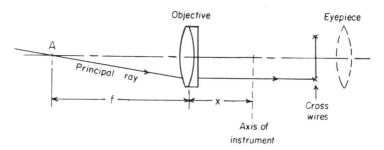

FIG. 13.6. A simple stadiameter.

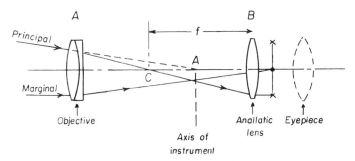

FIG. 13.7. Porro's anallatic stadiameter.

In 1850 Porro suggested that the instrument constant could be eliminated by the use of an *anallatic lens* (see Fig. 13.7). The entering principal ray is directed toward the anallatic point A, which is now located on the vertical axis of the instrument. After refraction at the objective lens, the ray crosses the axis at C, which is the anterior focus of the anallatic lens located close to the image. The cross-wires are separated by 1/100 of the overall focal length of the system, as usual. However, many workers feel that the simpler instrument is perfectly satisfactory and that the addition of the instrument constant to the reading does not cause any problem.

B. A Laser Rangefinder

The flash of light from a solid-state laser is so bright that it is possible to detect the reflection from a distant building or other target, and to measure the time interval between the emission of the flash and the reception of the return signal. This time is extremely short, of the dimension of microseconds or even nanoseconds, but it is possible to make this measurement electronically and thus develop a laser rangefinder. For instance, if the target distance is 200 m, the time of flight of the light over the go-and-return distance of 400 m is only 1.33 μsec because light travels at 3×10^8 m/sec. If this time can be measured to an accuracy of 1 nsec, the range will be known to ± 15 cm, or 6 in. This is a constant error for all ranges within the capacity of the instrument.

Optically, a laser rangefinder consists of three basic units: a solid-state laser to emit the flash, a viewfinder system to observe the object to be ranged, and a photoelectric receiver, located behind a pinhole at the focus of a lens, to detect the return flash from the object. The three units must, of course, be boresighted together, and the reticle in the viewfinder telescope must indicate precisely the position of the pinhole in front of the detector. The laser flash illuminates a wide expanse of territory, but the instrument will indicate the range of only the object whose image falls into the pinhole.

The military regard the laser rangefinder as an "active" system because it emits a flash of infrared radiation that could reveal the presence of the rangefinder to an observant enemy. Therefore, they often prefer the old-fashioned "passive" type of rangefinder in spite of its greater complexity and the need for a trained observer.

C. Military, or Self-Contained-Base Rangefinders

The general arrangement of a military rangefinder is indicated schematically in Fig. 13.8. At the two ends of a base line, which may be between 1 and 20 m in length, are 90° reflectors that form images of the distant target in the eyepiece. At the infinity setting the entering beams are parallel, and the system is arranged so that the two images fall together in the eyepiece focal plane and appear to be in coincidence. Generally, one image is made to fall above the other, with a horizontal dividing line between them. Sometimes the upper image from one beam is inverted so that it appears to be a mirror image of the lower image from the other beam. The two end mirrors are actually

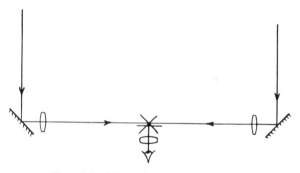

Fig. 13.8. Schematic military rangefinder.

II. RANGEFINDERS

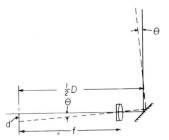

FIG. 13.9. Right-hand side of rangefinder when focused on a near object.

pentaprisms or pentamirrors which effect a constant 90° deviation of the light even if they are seriously displaced from their correct positions. The two mirrors indicated diagrammatically in the eyepiece focal plane are actually parts of a complex coincidence-prism assembly.

When the instrument is focused on a near object, the left-hand beam is unchanged, but the right-hand beam is tilted through an angle θ (Fig. 13.9), which is such that

$$\theta = D/R \text{ rad} = 3438 \, D/R \text{ arc-min},$$

where D is the base length and R the distance of the object being measured. The image formed by the right-hand half of the instrument is displaced laterally in the focal plane by a distance $d = f\theta$, and the observer brings the two images back into coincidence either by changing the slope angle of the beam or by displacing the image laterally. Both these possible procedures have been used in actual rangefinders. The main difficulty is the smallness of the angle θ, which may be only a few arc-min, and the extreme accuracy with which it must be measured. The tolerance on θ is a few arc-sec. at the most, and the tolerance on the lateral image displacement d is only a few thousandths of an inch.

The angle θ can be measured by

(a) tilting the end mirror through an angle $\theta/2$;
(b) tilting a small-angle glass wedge in a parallel beam;
(c) rotating a pair of small-angle wedges about the lens axis through equal angles in opposite directions.

Alternatively, the lateral image displacement d can be measured directly by

(d) moving the objective lens laterally through a distance d;
(e) tilting a parallel glass plate in a converging beam;
(f) sliding a glass wedge along a converging beam.

The two simple procedures (a) and (d) are far too imprecise for military applicatons, although they have been employed in camera rangefinders where the value of θ is large and the required precision is low.

Methods (b) and (e) are quite practical, and they have the advantage that they are most sensitive when θ is the smallest, i.e., for very distant objects.

Method (b). Suppose, for example, that the rangefinder base is 3 m long; then the angle θ to be compensated is given by the following tabulation.

Range (m)	10,000	5,000	1,000	500
Angle θ (arc-min)	1.03	2.06	10.31	20.63

Assuming that the tilting prism is made of glass having a refractive index equal to 1.523, then the necessary tilt of the prism from its initial symmetrical position to cause a deviation of 22 arc-min is shown (approximately) in the following tabulation.

Prism angle (deg)	1	2	3
Maximum tilt (deg)	45	36	31

Method (e). Figure 13.10 shows that when a converging beam passes through a tilted parallel glass plate, the image will be displaced laterally by

$$S = t\left[\frac{\sin(I - I')}{\cos I'}\right],$$

where t is the thickness of the plate. Assuming that the objective lens has a focal length of 50 cm and the refractive index of the plate is 1.523, the lateral image displacement corresponding to an object at 500-m distance would be about 3 mm, obtainable in any of the ways shown in the following tabulation.

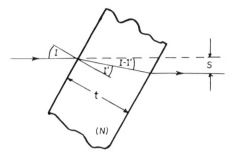

FIG. 13.10. Lateral image displacement caused by a tilted glass plate in a converging beam.

Plate thickness (mm)	15	20	25
Tilt angle (deg)	30	23	19

Method (f). Method (f) is shown in Fig. 13.11. A glass wedge that deviates the light through an angle β is moved along the converging beam from the right-hand objective toward the image. The angular deviation of the light remains constant but the lateral displacement of the image in the focal plane diminishes in a linear manner as the wedge is moved along from the lens to the image. Thus, for example, if the rangefinder base is 1 m and the focal length of the objective lens is 40 cm, then to cover ranges from 50 to 500 m the angular deviation of the wedge must be about $\beta = 78$ arc-min. Assuming these conditions, the lateral displacement of the image is shown in Table I. This type of construction has been used extensively in 1-m hand-held rangefinders. The range scale is attached to and moves with the sliding wedge in front of a magnifying lens placed conveniently close to the main eyepiece.

FIG. 13.11. The sliding wedge method for displacing an image laterally.

TABLE I

TYPICAL DATA FOR A ONE-METER RANGEFINDER

Range (m)	Angular displacement θ (min)	Lateral shift of image = d (cm)	Distance from wedge to image (cm)
500	6.88	0.08	3.50
200	17.19	0.20	8.75
100	34.38	0.40	17.50
50	68.75	0.80	35.00

Method (c). If the method of oppositely rotating wedges is used a pair of glass wedges, each of which deflects the beam through an angle ϕ, is placed in series in a parallel beam. Looking end-on at the wedges, we may represent each angular deflection as a vector (Fig. 13.12). Thus, when the wedges are rotated about the lens axis by equal amounts in opposite directions, the combined deflection remains in a constant orientation but varies in magnitude from 2ϕ to the left to 2ϕ to the right, or 4ϕ altogether. Thus a maximum deflection of 1° can be obtained by the use of two wedges, each of which provides a deviation of 15 arc-min. The wedges are rotated by a gearing similar to the differential gear in an automobile, and of course the range scale is highly nonlinear. This arrangement is used in some of the most precise military rangefinders.

D. STEREOSCOPIC

If the two images in a self-contained-base rangefinder are brought to the two eyes separately instead of being combined in a single eyepiece, a stereoscopic rangefinder is formed. A reticle in the focal plane of each

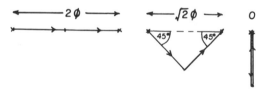

FIG. 13.12. Vector representation of a pair of oppositely rotating wedges.

eyepiece carries a small dot, or *wandering mark,* which appears to move toward or away from the target when the range setting is made. Of course, it is actually the image that moves and the wandering mark that remains fixed, but the observer has the feeling that the mark is moving toward or away from the object. With a little practice, an observer having a good stereoscopic sense can make the dot fall precisely on a given target, and he can then read off the range in the usual way. Theoretically, a stereoscopic and a coincidence rangefinder have about the same ultimate precision, but in some cases the stereoscopic instrument is found to be easier to use. Not everyone has a well-developed stereoscopic sense, however, and in such cases the coincidence type is preferred.

III. AN AXICON

As its name implies, an *axicon* is any device that forms an image of a point source lying along the axis of the device. A typical axicon can be made from a weak lens having a plane surface on one side and a slightly conical surface on the other, as shown in Fig. 13.13. Light from a distant point source passing through any one zone of the cone lens emerges at a fixed angle on the other side and crosses the axis at the focal point of the zone. The line of foci, therefore, starts at the tip of the cone and extends for a long distance along the axis. The actual length of the axicon image depends, of course, on the angle of the cone.

At any beam section such as *P*, there will be a small sharp dot of light on the axis surrounded by a large circle of light. Starting from the tip of the cone and moving along the beam, the central dot will grow brighter because the area of the zone increases, and the surrounding circle of light will become smaller, but the appearance of the central dot image will remain the same because the ray slope U' is constant for all zones. If the object is not on the lens axis, the image in the middle of the beam will show four cusps instead of being a dot.

It is not necessary to use a cone lens to make an axicon, and any lens

FIG. 13.13. A typical axicon.

having a sufficiently large amount of spherical aberration may be used. In that case, the emerging ray slope U' will not remain constant, and the appearance of the central dot image will change from point to point along the axis.

An axicon has been used to provide a line of light to test the straightness of a lathe bed or the alignment of the bearings for the shaft of a ship propeller, but today the beam from a small gas laser is generally used instead.

A variety of axicon-like mirror systems has been developed for special purposes, with names like "reflaxicon" and "waxicon." A summary of these has been given by Ferguson.[3]

---

[3] T. R. Ferguson et al, "Conical element nomenclature, use, and metrology." *Opt. Eng.* **21**, 959 (1982).

CHAPTER 14

Mirror Imaging Systems

Reflective optical systems in which a concave mirror, or a combination of lenses and mirrors, is used to form an image are becoming increasingly common. The chief reason for this is the weight and cost of the piece of optical glass required to make a large lens. Consequently, the limit in size for a lens system is about a 10- or 12-in. diameter, although plastic lenses can be made considerably larger. A mirror, of course, can be made of any suitable material, including metals, because the light does not penetrate the material but is reflected from its surface. Thus there is no limit to the size of a mirror, which may be as small as a few millimeters in diameter or as large as 20 ft. A mirror also possesses no chromatic aberrations of any kind, so it can be focused in the visible and used in the UV or IR with no change in focus and no variation in the aberration correction.

A mirror is much weaker than a lens, and the aberrations of a mirror are likely to be as small as one-quarter of the aberrations of a comparable lens. The ancient term "catoptric" refers to an all-mirror system, whereas the term "catadioptric" refers to a system containing both mirrors and lenses.

I. SINGLE-MIRROR SYSTEMS

A. SPHERICAL

The most perfect optical system, which has no aberrations of any kind, is a spherical mirror with an object point at its center of curvature. The image is also at the center of curvature. The object and image can be separated by a beam splitter, as in Dyson's device for increasing the working distance of a microscope (Fig. 14.1a). Light from the object, located at the center of curvature of the mirror, passes through the thin plane beam splitter, is reflected at the spherical mirror, and is

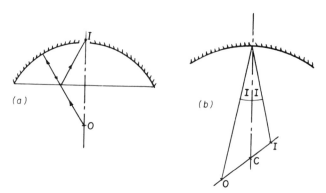

FIG. 14.1. Images formed by a spherical mirror.

then reflected at the beam splitter to a small hole in the vertex of the spherical mirror, where an image is formed that is identical with the original object. The image in this system is much dimmer than the object because the optimum reflectivity of the beam splitter is 50%, in which case only 25% of the light reaches the final image.

It should be noted that if the object is moved away from the center of curvature of a spherical mirror, the location of the image point can be found immediately. The object point, image point, and center of curvature of the mirror lie along a straight line, and the angles of incidence and reflection of a ray at the mirror are equal, so the intersection of these two loci locates the image (Fig. 14.1b).

If a single spherical mirror is used to form an image, the principal point is at the mirror and the nodal point is at the center of curvature, the focal length being half the radius of curvature. Furthermore, when the object is at infinity, a sizable amount of spherical aberration arises, and several methods have been proposed to remove it.

A cat-eye reflector. A familiar example of a combination of a lens and a mirror is the common cat-eye reflector used on highway signs (Fig. 14.2). This consists of a solid piece of glass with an elliptic refracting surface in front and a spherical mirror in back. Parallel rays from a distant automobile headlight enter the front of the unit and are sharply focused on the mirror, from which the light is autocollimated back through the front surface toward the source. The whole unit thus appears brilliantly lit to the driver of the car. However, as in the case of the cube-corner plate (Section III of Chapter 9), a perfect unit would

I. SINGLE-MIRROR SYSTEMS

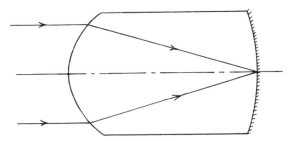

FIG. 14.2. A cat-eye reflector.

not be useful because all the light would be sent back into the headlamp and not toward the driver of the car, but the small random errors in the manufacturing process introduce just enough spread in the reflected beam to make the device usable.

The Mangin mirror. The French engineer A. Mangin, in 1876, proposed that the familiar parabolic mirror in a searchlight could be replaced by a spherical mirror formed by silvering the convex side of a negative meniscus lens, the overcorrected spherical aberration of the lens being just sufficient to offset the undercorrected aberration of the spherical mirror. A Mangin mirror is too heavy to be used in a large searchlight, but it was used for many years as the reflector in acetylene automobile headlights.

The Schmidt camera. In 1932 the Estonian optician B. Schmidt suggested that a concave spherical mirror could be used in astronomical telescopes by mounting an aspheric corrector plate in the entering beam, the asphere being calculated to correct the spherical aberrration of the spherical mirror (Fig. 14.3). Furthermore, by placing the corrector plate at the center of curvature of the mirror, he was able to eliminate both the coma and the astigmatism of the system and thus

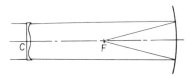

FIG. 14.3. An $f/2$ Schmidt camera.

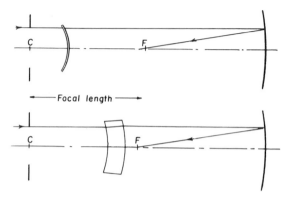

FIG. 14.4. Two possible f/3 Bouwers–Maksutov monocentric systems.

cover a field as wide as 5 or 6° from the axis. The focal surface is concentric with the mirror, and slightly curved photographic plates are used.

The Bouwers–Maksutov system. The aspheric plate in the Schmidt system is difficult to make, and it occurred almost simultaneously to A. Bouwers and D. Maksutov that a thick concentric meniscus lens, mounted concentrically with a spherical mirror, would introduce a sufficient amount of overcorrected spherical aberration to offset the undercorrection of the mirror (Fig. 14.4). If the lens is thin, it must be placed close to the stop so that it is strongly meniscus shaped, whereas if it is thick it must be moved further from the stop and be less strongly meniscus. In a monocentric system such as this, the angular field is, of course, unlimited, and the focal surface is also concentric with the mirror. In large sizes the weight of the corrector lens is a serious disadvantage, but in small sizes this is no problem. The nodal points are at the common center, and the focal length is half the radius of curvature of the mirror.

B. Elliptical

The equation of a conic section with origin at the vertex is

$$Y^2 + Z^2(1 - e^2) = 2Z/c,$$

I. SINGLE-MIRROR SYSTEMS

where c is the vertex curvature and e the eccentricity. Hence, solving for Z gives

$$Z = \frac{cY^2}{1 + \sqrt{1 - c^2 Y^2 (1 - e^2)}}.$$

When $e = 0$ the curve becomes a circle, and when $e = 1$ it becomes a parabola. If e lies between 0 and 1, the curve is an ellipse, and if $e > 1$, the curve is a hyperbola.

An ellipse has a major and minor semiaxis (Fig. 14.5), which are usually denoted by a and b. The two vertex radii are b^2/a and a^2/b, respectively. The eccentricity is given by $e = \sqrt{a^2 - b^2}/a$. Hence

$$a = \frac{1}{c(1 - e^2)}, \quad b = a\sqrt{1 - e^2} = \sqrt{\frac{a}{c}}, \quad \text{and} \quad c = \frac{a}{b^2}.$$

An ellipse has two foci, F_1 and F_2, and the focal lengths from the vertex are

$$f_1 = a(1 - e) \quad \text{and} \quad f_2 = a(1 + e) = 2a - f_1.$$

These foci have the property that any ray starting out from one focus passes through the other focus after reflection at the ellipse. Thus, knowing the two focal lengths, we find that

$$e = \frac{f_2 - f_1}{f_2 + f_1}, \quad a = \frac{(f_1 + f_2)}{2}, \quad b = \sqrt{f_1 f_2}, \quad \text{and} \quad c = \frac{f_1 + f_2}{2 f_1 f_2}.$$

The image magnification along any ray reflected from F_1 to F_2 via a point P on the ellipse is given by $m = PF_2/PF_1$. Since this value changes from point to point along the ellipse, it indicates the presence of a considerable amount of coma in the image. For this reason an elliptical mirror should be used only when the object is very close to the axis.

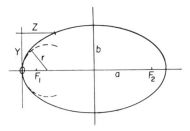

FIG. 14.5. An elliptic mirror ($e = 0.8$).

C. Parabolic

The equation of a parabola, with $e = 1$, is simply

$$Z = Y^2/2r = \tfrac{1}{2}cY^2,$$

where r is the vertex radius and c the vertex curvature. A parabolic mirror has a focus located midway between the vertex and the center of curvature at the vertex, so that paraxially $f' = r/2$. It has the property that all rays entering the mirror parallel to the axis are reflected through this focus. However, as the focal length of each ray is equal to the distance from the point of incidence to the focus, the focal length keeps increasing for rays progressively further out from the axis (Fig. 14.6), leading to the presence of a large amount of coma in off-axis image points. Like the ellipse, a parabolic mirror should be used only when the image is very close to the axis.

An off-axis parabolic mirror. The obstruction of the entering beam caused by the image receiver can be eliminated for a parabolic mirror by using only part of the mirror with the parabolic axis lying outside the mirror aperture (Fig. 14.7). This arrangement is called an off-axis parabola, although a more accurate name would be an off-aperture mirror, because the image must lie on the mirror axis if coma is to be

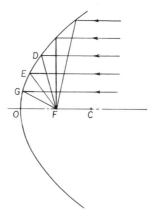

FIG. 14.6. A parabolic mirror. The focal lengths of the mirror for the various zones are *DF*, *EF*, etc. The paraxial focal length is *OF*.

I. SINGLE-MIRROR SYSTEMS

FIG. 14.7. An off-axis parabolic mirror.

avoided. Thus, three or more off-axis mirrors can be cut from one large parabolic mirror, as indicated in Fig. 14.8. Alternatively, a group of separate small mirrors could be polished at one time by mounting them in a ring around the axis of the polishing block.

There are many uses for an off-axis parabolic mirror, two of which are given in the following list.

(a) *A schlieren system.* An imaging system for observing the schlieren, or compression waves, in a wind tunnel involves two identical off-axis parabolic mirrors, as indicated in Fig. 14.9. An illuminated slit at S_1 is collimated by the first mirror to form a parallel beam in the tunnel and then reimaged by the second mirror at S_2 where an opaque bar blocks off the direct light. Any variation in the refractive index of the air in the tunnel causes the light to be slightly deflected so that it misses the opaque bar and becomes visible to the eye at E. If the object in the tunnel is roughly at the focus of the second mirror, the image of the striations will be at infinity, and a camera can be substituted for the eye without any focusing problems.

It should be noted that the axes of the parabolic mirrors pass through the slits S_1 and S_2 and lie parallel to the parallel light in the tunnel. The two mirrors must be in the Z configuration shown in Fig. 14.10b so that the coma of one mirror will be cancelled by that of the other. This constitutes a symmetrical system, referred to in Section IV of Chapter 2 when describing the various aberrations. If the mirrors were arranged

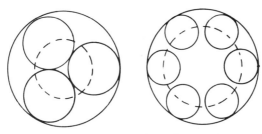

FIG. 14.8. Cutting several off-axis mirrors from one large paraboloid.

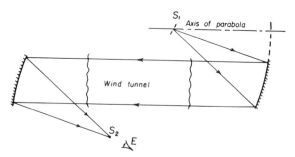

FIG. 14.9. A schlieren system for a wind tunnel.

as in Fig. 14.10a, the two comas would add, giving a useless image at F_2.

(b) *The Wadsworth Mirror Monochromator.* The Wadsworth monochromator, described in Section I of Chapter 16, is often constructed with concave mirrors instead of lenses so that the image remains in focus for all wavelengths. A pair of identical off-axis parabolic mirrors is used, and in the Z arrangement shown in Fig. 16.6, the coma of one mirror cancels that of the other, giving a sharp image of the slit in the spectrum lines at the second slit.

(c) *A head-up display.* A head-up display, or HUD, is a means for providing an airplane pilot with alphanumeric or other data from a cathode-ray display without the necessity for him to lower his gaze from the distant scene.

The display is projected onto a large transparent parabolic mirror M located between the windshield and the pilot's head, as indicated in Fig. 14.11. The material to be displayed is formed on the face of a small CRT C located above the pilot's head, and it is relayed by a lens L to

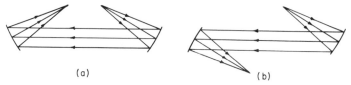

FIG. 14.10. A pair of off-axis parabolic mirrors, where (a) the comas add; (b) the comas cancel.

II. TWO-MIRROR SYSTEMS

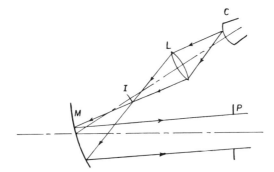

FIG. 14.11. A typical head-up display.

the focus of the concave mirror at I, from which it is projected as a collimated beam to the pilot's eyes. The exit pupil P of the system, within which his eyes must be located, is an image of the relay-lens aperture formed by the parabolic mirror, and it should be as large as possible to give the pilot the maximum freedom in the location of his head. There is inevitably some degree of distortion in the displayed image, and when the pilot moves his head sideways the image appears to "squirm." Attempts have been made to use some other shape for the mirror but without much success.

D. Hyperbolic

The equation of a hyperbola is the same as that of an ellipse except that the eccentricity is now greater than 1.0 instead of less. A convex hyperboloid resembles an ellipsoid except that now one of the two foci is virtual; this arrangement is employed in the Cassegrain telescope. A concave hyperbolic mirror is used in the Ritchey–Chrétien telescope system (see Section III).

II. TWO-MIRROR SYSTEMS

Since the time of Newton, reflecting telescopes have usually contained two mirrors—a large concave *primary* mirror to receive the parallel light from a distant object and a smaller *secondary* mirror. For a long time the preferred form was the Gregorian, in which the primary was a paraboloid and the secondary a concave ellipsoid; this yielded an erect image that was necessary for terrestrial observation. Much later the Cassegrain form became universal for astronomical purposes because of its greater compactness and smaller obstruction. In this case the secondary was a convex hyperboloid, an exceedingly difficult form to fabricate.

A comparison of the three basic mirror telescope systems (all at the same focal length) is shown in Fig. 14.12. In part (a) is the Newtonian form, with a parabolic mirror and a small diagonal flat to place the image outside the tube. The obstruction caused by the diagonal is minimal, but the system is very long and it is used mainly by amateurs. The Gregorian is shown in (b), where the relative apertures of the two mirrors P and S are $f/2$ and $f/1.5$, respectively, and the diameter obstruction caused by the secondary is 50%. Lastly, with the Cassegrain form (c), the relative apertures of the mirrors shown are $f/2$ and $f/3$, respectively, leading to an obstruction of only 25%. By temporarily removing the secondary mirror, observations can be made at the prime focus F, where the high relative aperture is useful for photographing faint extended objects in the sky.

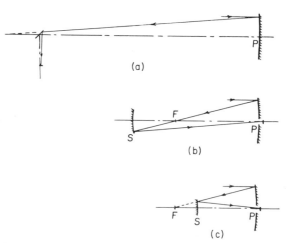

FIG. 14.12. Astronomical reflecting telescopes: (a) Newtonian, (b) Gregorian, (c) Cassegrain.

II. TWO-MIRROR SYSTEMS

A. THE DALL–KIRKHAM TELESCOPE

Amateur telescope makers have great difficulty in fabricating a telescope of the Cassegrain type, mainly because of the impossibility of properly testing a convex hyperboloid mirror. For this reason the arrangement suggested by Dall and Kirkham is often preferred. In this system the primary mirror is elliptical, which can be conveniently tested in the workshop by placing a small source at one focus and a knife-edge at the other, and the convex secondary is spherical, which can be tested by Newton's rings in the ordinary way. The only objection to this system is the greatly increased coma. After all, as Ritchey and Chrétien pointed out, the edge curvature of a paraboloid must be weakened to correct the coma, but in the Dall–Kirkham form the edge of the primary is steeper than in a parabola.

B. LASER-BEAM EXPANDER

It is possible to construct a Cassegrain laser-beam expander from two parabolic mirrors, one small and convex and the other large and concave (Fig. 14.13). For example, if the first mirror has a radius of 4.0, separated by a distance of 18 from a second mirror of radius 10.0, then we have an afocal system with a magnifying power of 10. Since a laser beam has a Gaussian cross section, it has been suggested that aspheric forms differing from a paraboloid be used to equalize the beam intensity across its width. It should be noted that with this arrangement the central one-tenth of the beam is missing.

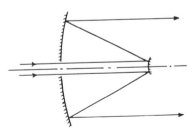

FIG. 14.13. A 10× Cassegrain-type laser-beam expander.

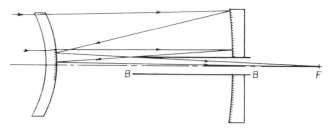

FIG. 14.14. The Questar optical system. The tube *BB* is a baffle to prevent unwanted light from getting to the image.

C. BAFFLES

When a Cassegrain mirror system is used for terrestrial purposes, it is possible for light to pass directly from surrounding objects to the image without being reflected at the two mirrors. To prevent this from occurring, it is necessary to install suitable baffles that will obstruct unwanted light without cutting into the useful beams. A typical arrangement of baffles for this purpose is indicated in Fig. 14.14, which shows the baffle system used in the Questar telescope. It is virtually impossible to baffle a simple Newtonian telescope, where the eyepiece is at the front of the tube, but the other types of telescope can be baffled if care is taken in the mechanical design.

The Questar telescope is an example of a type of modified Cassegrain telescope in which both mirrors are spherical and the inevitable spherical aberration is corrected by the insertion of a thick meniscus lens in the entering light, in the manner suggested by Bouwers and Maksutov for a single spherical mirror. The secondary mirror is conveniently formed by evaporating a disk of aluminum in the middle of the rear face of the correcting lens.

III. COMA CORRECTION

In a parabolic or Cassegrain telescope system, there will be a large amount of coma appearing in images that lie a short distance to one

III. COMA CORRECTION

side of the axis. The magnitude of this coma is found to be

$$\text{coma} = h'/16(F\text{-number})^2,$$

where h' is the height of the image above the axis. Since the coma depends only on the F-number of the system and not on its focal length, it is as bad with the 200-in. telescope as with a 20-in. instrument having the same F-number. Thus the size of the coma-free field is surprisingly small for an $f/3.5$ telescope, but it is quite large at low apertures such as $f/16$.

Various suggestions have been made for the correction of this serious amount of coma. In 1935 F. E. Ross suggested inserting an airspaced doublet lens of zero power into the telescope tube some little distance upstream from the image. In this way he could correct the coma and flatten the field, but at the expense of increased spherical aberration. Later designs by others have made it possible to correct all aberrations in this way.

A better suggestion was made by G. W. Ritchey and H. Chrétien in the 1920s. As coma is caused by the fact that the focal length of the objective varies with aperture, they argued that since focal length is related to the emerging ray slope angle by

$$F' = Y/\sin U',$$

where the value of U' is a little too small for rays entering at a high value of Y, it might be possible to increase U' for high values of Y by slightly flattening the rim of the primary mirror by making it a weak hyperboloid instead of a paraboloid and slightly flattening the rim of the secondary mirror also (Fig. 14.15). In this way they were able to make the focal length of all rays the same and thus remove the coma.

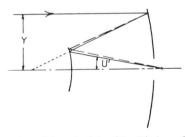

FIG. 14.15. Illustration of the principle of the Ritchey–Chrétien telescope.

This type of construction is becoming standard for all large telescopes today.

A. The Schwarzschild Microscope Objective

Karl Schwarzschild, in about 1904, discovered that a monocentric pair of spherical mirrors, suitably spaced, is automatically corrected for spherical aberration and coma. It has the usual field curvature characteristic of all monocentric systems. In this mirror microscope, shown to scale in Fig. 14.16, the light first strikes the large concave mirror and then the small convex mirror, from which it proceeds to the eyepiece of the instrument.

Interestingly enough, the Cassegrain and Schwarzschild systems are the analogs of the telephoto and reversed telephoto lens arrangements, in which we regard a concave mirror as the analog of a positive lens and a convex mirror as the analog of a negative lens. Thus, if we require a system having a total length shorter than the focal length, we can use either a telephoto lens or a Cassegrain mirror; conversely, if we require a long back focus with a short focal length, we can select a reversed telephoto lens or a Schwarzschild mirror arrangement.

As a microscope objective, the real advantage of the Schwarzschild system is that it has no chromatic aberration, so it can be focused in the visible and then used in the UV or IR region of the spectrum with no change in either the focus or the aberration correction. Some manufacturers have added lenses to the Schwarzschild system to increase the numerical aperture, but this, of course, destroys the principal advantage of the system.

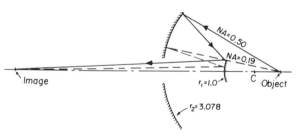

FIG. 14.16. A 10× Schwarzschild microscope objective.

IV. OBSTRUCTION

One of the problems in using a mirror to form an image is that the image receiver, be it a film, an eyepiece, a television tube, or a photocell, is located within the entering beam, and it necessarily obstructs some of the entering radiation. However, this is not as bad as it might seem, for even if the diameter of the obstruction were as large as half the diameter of the mirror, it would only block out one-quarter of the light, which amounts to 0.42 of a stop, so that if the mirror had an aperture of $f/4$, the 50% obstruction would reduce it only to $f/4.6$. However, a more serious effect of the obstruction is that the first bright ring of the Airy disk is considerably brightened and the MTF at low frequencies is noticeably lower than for the unobstructed aperture.

It is shown in books on physical optics[1] that the expression for the *amplitude* of the light at a radial distance r out from the lens axis, in the absence of all aberrations, is given by

$$2J_1(w)/w,$$

where $w = 2\pi a r/\lambda l$. Here a is the radius of the lens aperture and l the distance from the lens to the image. The *intensity* of the light at the point r is given by the square of the amplitude, as stated in Section III of Chapter 2. The factor 2 is merely a normalizing factor to make the central intensity equal to 1.0.

When there is a central obstruction having a diameter equal to a fraction ϵ of the lens aperture, we merely subtract the amplitude corresponding to the obstruction from that of the open aperture, remembering that the amount of light in the obstruction is less than that in the open aperture by a factor of ϵ^2. Hence, the amplitude with the central obstruction in place becomes

$$\frac{2J_1(w)}{w} - \epsilon^2 \left[\frac{2J_1(\epsilon w)}{\epsilon w} \right],$$

and the intensity is the square of this amount. The term w refers to the value of $2\pi a r/\lambda l$ for the open aperture.

As an example, we will calculate the distribution of light in the Airy disk for a 50% obstruction, noting that, as usual, the intensity is the square of the amplitude. Then, since ϵ is equal to 0.50, we find that the

---

[1] M. Born and E. Wolf, "Principles of Optics," 2nd ed., p. 416. Pergamon, Oxford, 1964; also Macmillan, New York, 1964.

intensity is

$$I = \left[\frac{2J_1(w) - J_1(\tfrac{1}{2}w)}{w}\right]^2,$$

and the normalizing factor to yield unit intensity at the central spot becomes $16/9 = 1.7778$.

Three graphs are shown in Fig. 14.17. They represent (a) a single marginal zone of the lens with $\epsilon = 1$ (this is merely a graph of J_0^2); (b) the usual graph for the open area of the lens; and (c) the graph for the 50% obstruction. It is not difficult to plot these graphs if tables of Bessel functions are available.

Notice that for a single narrow zone of the lens the central spot is fairly narrow but the rings are very bright and extend a long way out from the axis. However, as more and more of the aperture is exposed, the central spot widens slightly but the rings rapidly become fainter, and for the fully open aperture, the rings have almost disappeared.

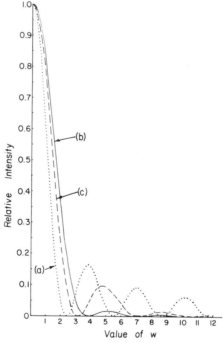

FIG. 14.17. Airy disk with a central obstruction.

IV. OBSTRUCTION

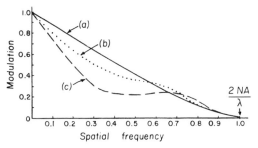

FIG. 14.18. Normalized MTF curves with a central obstruction.

The effect of a central obstruction on the MTF of the system is also of interest. Three cases are shown in Fig. 14.18. Curve (a) is the MTF for an open circular aperture, the resolving limit occurring when the spatial frequency is equal to $2\,\text{NA}/\lambda$. Curve (b) represents the situation with a 25% central obstruction, and curve (c) is the case of a 50% central obstruction. The drop in contrast at low spatial frequencies is, of course, due to the brighter rings surrounding the image of a point, but the slight increase in MTF near the limit of resolution is caused by the central spot being a little narrower for a central obstruction than for an open circular aperture.

CHAPTER 15

Photographic Optics

I. PERSPECTIVE EFFECTS IN PHOTOGRAPHY

A picture, whether printed on paper or projected on a screen, is a two-dimensional representation of a three-dimensional object, and it will "look right" only if the viewer is located at such a position that objects in the picture subtend the same angles at his eye as the original objects subtended at the camera. This point is known as the *center of perspective* of the scene.

Consider the case of a portrait taken with a short-focus lens. To make the person appear reasonably large on the film, the photographer moves up close to the subject, whose nose is then appreciably closer to the camera than his ears. As a result, his nose will appear unduly large in the picture and his ears unduly small. This effect is commonly called *wide-angle distortion* because the short-focus lens is used to cover a wide angle of field. However, this distortion will disappear if the picture is viewed from its correct center of perspective, which may be uncomfortably close to the picture.

In the opposite case, when a long-focus camera lens is used, the photographer must stand at a considerable distance from the subject, and then close objects will appear too small when compared with distant objects. This picture must be viewed from a considerable distance if it is to appear natural.

It is worth noting that an 8×10-in. print made from a 35-mm negative and viewed from a distance of 15 in. subtends approximately the same angle as the original film subtended at a lens of 45-mm focal length. Similarly, a color slide projected on a screen and viewed from a distance of three times the width of the image subtends the same angle as at a camera lens of approximately a 100-mm focal length. If realism is important, the photographer should take care to use the right lens for whatever viewing conditions will be used for the final picture. If a color slide is taken with a 50-mm lens and projected with a lens of 5-in. focal length, the center of perspective in the

II. FOCUSING ON A NEAR OBJECT

audience will be at 40% of the distance of the screen from the projector, or a little closer than the midpoint of the audience.

II. FOCUSING ON A NEAR OBJECT

Most 35-mm cameras today are equipped with lenses that can be focused on an object lying anywhere between infinity and about 10 focal lengths distant, representing an image magnification ranging from zero to about 0.1×. Thus a typical 50-mm lens can be focused down to 0.5 m (20 in.) and a 100-mm lens to about 1 m (40 in.). For mechanical reasons, many large long-focus lenses can be moved through only about 15 to 20 mm, representing a closest object at a distance of perhaps 15 to 20 ft.

Several possible procedures are available for focusing on closer objects to obtain a higher image magnification. Some of these are discussed in the following paragraphs.

A. By Use of Diopter Attachments

Adding a thin meniscus lens of 1-, 2-, or 3-diopters power in front of the camera lens shortens the focusing distance appreciably. We can calculate the new object distance by the rule that

$$\frac{1}{\text{focusing distance}} = \frac{\text{diopter power}}{39.4} + \frac{1}{\text{focus scale setting}},$$

all distances being in inches. This formula was used to calculate the values in Table I. (This is a good place to warn the user that while the 1D and 2D attachments are generally quite satisfactory, the 3D attachment may show signs of distortion and lateral color, even when the main lens is stopped down considerably.)

B. By Means of an Extension Tube

Many manufacturers provide a series of extension tubes of differing thickness that can be inserted between a camera and its lens. The

TABLE I

FOCUSING DISTANCE USING DIOPTER LENSES

Focus scale setting (ft.)	Actual focusing distance with diopter lens (in.)		
	1D	2D	3D
∞	39.4	19.7	13.1
50	36.9	19.0	12.8
25	34.8	18.5	12.6
15	32.3	17.7	12.2
10	29.6	16.9	11.8
8	27.9	16.3	11.5
6	25.5	15.5	11.1
5	23.8	14.8	10.8
4	21.6	14.0	10.3
3	18.8	12.7	9.6
2	14.9	10.8	8.5

object distance p is now related to the focal length f, the thickness of the extension tube E, and the focus scale setting s on the camera by

$$p = f\left[\frac{sf + E(s-f)}{f^2 + E(s-f)}\right] \quad \text{and} \quad s = f\left[\frac{pf - E(p-f)}{f^2 - E(p-f)}\right].$$

Thus, for example, suppose that $f = 50$ mm, $E = 15$ mm, and $s = \infty$. Then to calculate the object distance p, we must divide the numerator and denominator by s, giving

$$p = f + f^2/E = 217 \text{ mm } (8.5 \text{ in.}).$$

Again, if we wish to use a lens with a 1-ft focal length to focus on an object 10 ft away, using an extension tube 1 in. thick, we can determine the lens setting by the second formula, giving $s = 35.4$ ft.

Of course, it is perfectly possible to combine an extension tube with a diopter attachment. For example, a 100-mm lens set at 1-m focusing distance will focus sharply on an object at 1000 mm. Adding a 2D attachment permits focusing at 333 mm. With a 15-mm extension tube instead of the diopter attachment, the focusing distance becomes 480 mm; with both the extension tube and the 2D attachment, the focusing distance drops to 270 mm, giving an image magnification of 1/2.7, or 0.37×.

C. By Use of a Telenegative Attachment

Instead of using an empty extension tube, a much greater change in image size can be obtained by using a 2× or 3× telenegative attachment, (a so-called tele-extender,) between the lens and the camera. This attachment serves to double or triple the image size without altering in any way the previously determined object distance. By means of such an attachment, combined with a diopter attachment if necessary, it is possible to copy a small object at a magnification close to unity.

D. By Use of a Small Bellows

If you are using a single-lens reflex camera (SLR), it is most convenient to use a *macro* lens with a small bellows between the lens and the camera. This type of lens has been especially designed for use at a wide range of magnifications with no loss of definition, and by extending the bellows sufficiently it is possible to reach unit magnification of even a little greater. It should be noted that at magnifications greater than unity, the lens should be turned around with the normal rear element facing the near object.

E. By Means of a Macro Zoom Lens

Some zoom lenses are called *macro zooms* because, by turning a locking ring, you can change the relation between the moving elements in such a way that it is possible to focus on very close objects without loss of definition and without the need for front or rear attachments.

III. DEPTH OF FOCUS AND DEPTH OF FIELD

The two terms "depth of focus" and "depth of field" are liable to be confused, and they are generally defined in the following manner. Assuming that the lens is free from all aberrations, there will be a certain plane in the object space that is precisely conjugate to the film plane in the image space. This is known as the *focused plane* in the object. Because the observer's eyes have a limited acuity, there will be a small amount of tolerable blur in a photograph that the average observer will be unable to distinguish from sharp imagery. Thus a point object may be imaged as a small *circle of confusion* on the film before the observer can detect that the image is not perfectly sharp.

The *depth of focus* of a lens is the distance along the lens axis in the image space from the plane of sharp definition to the place where the image of a point source just reaches this permissible circle of confusion. Similarly, the *depth of field* of a lens is the corresponding distance of an object point from the focused plane in the object space before its image reaches the permissible circle of confusion in the film plane.

To make these matters more specific, it is clear that the depth of focus is obviously equal to the product of the diameter c' of the permissible circle of confusion on the film and the F-number N of the lens. This distance is the same whether the image is just within or just beyond the film plane.

In the object space things are more complicated because objects lying beyond the focused plane may at times extend to infinity before their image becomes significantly out of focus on the film, whereas near objects soon reach a limit to their acceptable distance from the focused plane. The situation is indicated in Fig. 15.1. This diagram is based on the fact that everything lying in the focused plane will be imaged sharply on the film, including the projection of a near or far object into that plane from the open aperture of the camera lens. Thus, if D_1 is the far depth of field, D_2 the near depth, c the diameter of the acceptable circle of confusion in the focused plane, and A the diameter of the lens aperture, then it is clear that

$$c/A = D_1/(s + D_1) = D_2/(s - D_2),$$

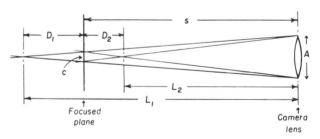

FIG. 15.1. Depth of field diagram.

where s is the distance of the focused plane from the lens. From these relations we see that the far and near depths of field are, respectively,

$$D_1 = cs/(A - c) \quad \text{and} \quad D_2 = cs/(A + c). \tag{1}$$

The far depth D_1 is clearly larger than the near depth D_2, and it would become infinite if $A = c$. The limiting distances from the camera are

$$L_1 = s + D_1 = As/(A - c) \quad \text{and} \quad L_2 = s - D_2 = As/(A + c). \tag{2}$$

We now consider what happens when a picture is viewed by an observer. We find that most people are incapable of distinguishing between a perfect point and a small circle of confusion that subtends about 2 arc-min (1 in 1700) at the eye. This is the subtense of 1 mm at a distance of 1.7 m (5½ ft). Three cases that arise for the observer are discussed in the following paragraphs.

Case 1. The eye is at the center of perspective. Having the eye at the center of perspective is the same as if the eye were at the center of perspective of the original scene, so $c = s/1700$. The limiting distances then become

$$L_1 = 1700As/(1700A - s) \quad \text{and} \quad L_2 = 1700As/(1700A + s).$$

Thus, in this case, the depth of field depends only on the distance of the subject and the linear diameter of the aperture of the camera lens. For example, a 2-in. lens at $f/2$ has a linear aperture of 1 in., so that if $s = 10$ ft, we find that

$$L_1 = 129 \text{ in.} = 10.8 \text{ ft} \quad \text{and} \quad L_2 = 112 \text{ in.} = 9.3 \text{ ft}.$$

Case 2. If the observer is not at the center of perspective. If the observer is not at the center of perspective, we must start with the observer and follow the sequence of events back to the camera and finally to the original scene. For example, suppose we have a 2-in. camera lens making a negative 1 × 1½ in. in size, which is enlarged onto a print of 4 × 6 in. This print is now viewed by an observer from a distance of 15 in. Assuming that his resolving power is 1/1700, the size of the acceptable circle of confusion on the print will be 15/1700 = 0.0088 in., and because the enlargement ratio was 4×, the circle of confusion on the film must have been $c' = 0.0022$ in. Projected back into the original object plane, we see that

$$c = c'(s/f) = 0.132 \quad \text{in.}$$

From Eq. (2) we see that if the lens aperture $A = 1$ in. ($f/2$) and $s = 120$ in. then

$$L_1 = 138 \text{ in.} = 11.5 \text{ ft}, \quad \text{and} \quad L_2 = 106 \text{ in.} = 8.8 \text{ ft.}$$

The depth of field is now nearly twice as great as it was in the first example because the observer of the print is about twice as far as he should be from the picture. The center of perspective of the picture is actually at 2 in. from the negative, or 8 in. from the print.

Case 3. A fixed circle of confusion on the film. In the case where c' has a fixed value, often assumed to be 1/30 mm, we have

$$c = c'(s/f) \quad \text{and} \quad A = f/N,$$

where N is the F-number of the lens. Then

$$L_1 = \frac{s}{1 - (Nsc'/f^2)} \quad \text{and} \quad L_2 = \frac{s}{1 + (Nsc'/f^2)}. \qquad (3)$$

These formulas assume that the viewer is located at a distance from the picture equal to $1700mc'$, where c' is the diameter of the circle of confusion on the negative and m the enlargement ratio between the negative and the final picture. Thus, for the example in case 2, the observer should be at a distance of $1700 \times 4 \times 1/30$ mm, which is about 9 in. from the print. This is very close to the center of perspective, but closer than most observers would care to be. After all, the 2-in. camera lens covers a field diagonal of about $\pm 24°$, and this is wider than the average viewer's ability to study a picture in comfort.

A. Effect of Lens Aberrations

Since spherical and chromatic aberrations in a lens have the effect of increasing the depth of focus, they also increase the depth of field. Indeed, the uncorrected lens in a simple box camera turns out to have a greater depth of field than it would have if the lens were well corrected, and such a camera need not be provided with any focusing arrangements. Also, as near objects form a larger image on the film than distant objects, it is easier to recognize a person standing too close to the camera than the same person standing too far away. This effect also tends to increase the near depth of field relative to the far depth, and as a result, the near and far depths in most cameras tend to be more nearly equal than these formulas would indicate.

IV. THE THEORY OF TILTED PLANES

It is well known that tilting a camera at the moment of exposure leads to some degree of *keystone distortion,* in which parallel lines in the subject appear as converging lines in the negative. If these converging lines are extended, they eventually meet at the *vanishing point V.* This type of distortion is independent of the lens, and it appears even in a pinhole camera; it is due solely to the plane of the film not being parallel to the plane of the subject.

A typical situation is indicated in Fig. 15.2. Here a tilted camera is shown photographing a vertical building AB. The vanishing point V is

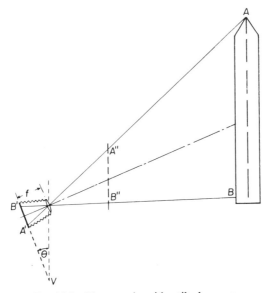

FIG. 15.2. Photography with a tilted camera.

found at the intersection of the plane of the film $A'B'$ and a vertical line drawn through the camera lens. The vanishing point is the conjugate of an infinitely distant point on the building if it were extended upward indefinitely.

It is not difficult to "rectify" converging parallels during printing by tilting the enlarger easel through the same angle θ by which the original camera was tilted. This process is equivalent to projecting the negative back onto a plane parallel to the original object but much closer to the camera, as at $A''B''$ in Fig. 15.2. Redrawn for the more familiar case of a vertical enlarger with a tilted easel, Fig. 15.2 becomes Fig. 15.3. The necessary tilt of the easel is found by joining the vanishing point V on the negative to the enlarging lens, and now the distance between the negative and the enlarging lens must be equal to the focal length f of the original camera lens. Hence, the focal length of the enlarger lens must be shorter than the focal length of the camera lens by an amount depending on the enlargement required. Thus, if the camera had a 4-in. lens and a 3× enlargement is desired, the enlarger must be equipped with a 3-in. lens. Failure to meet this condition results in some degree of anamorphic distortion, because the easel would not have the same slope as the original camera.

One further point should be mentioned. To obtain sharp imagery throughout the entire easel, it is necessary to tilt the enlarger lens slightly, so that the planes of the negative, lens, and easel all meet at the point S in accordance with the Scheimpflug condition(see Section V of Chapter 4). This tilt is not great, but it makes an enormous difference to the sharpness of the final print. As the point B'' is closer to the

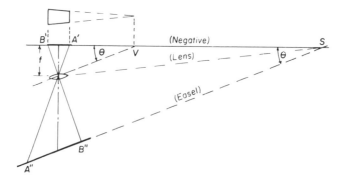

FIG. 15.3. Arrangement of an enlarger for "rectifying" a keystoned negative.

IV. THE THEORY OF TILTED PLANES

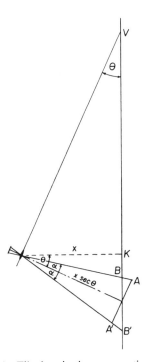

FIG. 15.4. Tilted projection on a vertical screen.

enlarging lens than the point A'', the exposure at B'' will be greater than at A'', and some degree of "dodging" must be employed to maintain an equal exposure over the whole picture. This requires care if it is to be successful.

Another example of keystone distortion is found when a motion-picture projector is placed high up in a theater and tilted downward toward the screen. This situation is shown in Fig. 15.4. The vanishing point is where a line through the projection lens parallel to the film intersects the plane of the screen, at the point V high up above the roof of the theater. The undistorted image is shown at AA' and the screen image at BB', so the screen image will be slightly keystoned, and also anamorphically stretched because BB' is longer than AA' in the vertical direction but the same size in the sideways direction. As an example, suppose that the length x of the theater is 100 ft, the downward tilt of the projector is 20°, and the half-field angle in the vertical direction is 10°. Then we see that

$$KB = 100 \tan 10° = 17.63 \quad \text{ft},$$
$$KB' = 100 \tan 30° = 57.74 \quad \text{ft}.$$

Hence, the vertical height of the screen image BB' is 40.11 ft. The height of the undistorted image is given by

$$AA' = 2 \times \sec 20° \tan 10° = 37.53 \quad \text{ft},$$

so that the anamorphic distortion in the vertical direction is 6.85%. In modern theaters, care is taken to ensure that the projector is nearly horizontal so that these distortions no longer appear.

V. SPECIAL PURPOSE LENSES

A. Telephoto Lenses

As mentioned in Section III of Chapter 5, a telephoto lens consists of a positive front component widely separated from a negative rear component, the purpose of this arrangement being to make the focal length greater than the total length from the front vertex to the focal plane. The negative rear component can be located anywhere between the front component and the focus, but its power is found to follow a parabolic variation with a minimum value when the negative component is located midway between the front component and the focus. The power of the positive component drops steadily from infinity when the two lenses are in contact, to a minimum when the negative lens is located at the image.

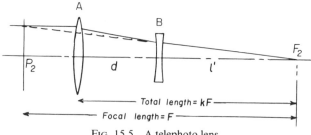

FIG. 15.5. A telephoto lens.

V. SPECIAL PURPOSE LENSES

Suppose the overall focal length is F and the *telephoto ratio*, or ratio of the total length to the focal length, is k, as indicated in Fig. 15.5. Then we know that

$$1/F = \phi_A + \phi_B - d\phi_A\phi_B, \qquad (4a)$$

and

$$l' = F(1 - d\phi_A).$$

But $l' + d = kF = F(1 - d\phi_A) + d$, from which

$$f_A = \frac{dF}{d + F(1 - k)}. \qquad (4b)$$

Substituting (4b) in (4a) gives

$$f_B = \frac{d(d - kF)}{F(1 - k)}. \qquad (4c)$$

To find the minimum value of f_B, we differentiate (4c), giving

$$\frac{\partial f_B}{\partial d} = \frac{2d - kF}{F(1 - k)} = 0, \quad \text{from which} \quad d = \tfrac{1}{2}kF.$$

As an example, suppose $F = 10$ and $k = 0.8$. Then

$$f_A = \frac{10d}{d + 2} \quad \text{and} \quad f_B = \frac{d^2 - 8d}{2}.$$

These values are plotted in Fig. 15.6.

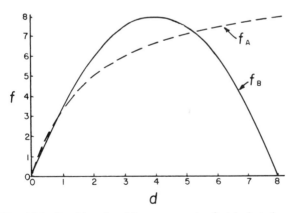

FIG. 15.6. Focal lengths of the components of a telephoto lens.

B. Reversed Telephoto Lenses

A lens of the reversed telephoto type has a large negative component in front and a much smaller positive component behind, just the opposite of the telephoto type of construction. It has the advantage of an exceptionally long back focus, usually longer than the focal length. Reversed telephoto lenses are universally employed for all the wide-angle lenses on an SLR camera because of the space needed to clear the 45° rocking mirror in the camera. Indeed, with a focal length of 18 mm and a back focus of 35 mm, the rear principal plane is about midway between the back of the lens and the film.

The front negative component of a reversed telephoto lens is invariably of a meniscus shape, and it serves to widen the angular field; the system as a whole tends to favor both a high aperture and a wide field. The only reasons why a reversed telephoto type is not used for every application are its large size and the high cost of the many lens elements necessary to correct all the aberrations.

C. Fish-Eye Lenses

The name "fish-eye lens" was coined by R. W. Wood, who realized that a fish looking upward in water would see a circular area of water surface which would contain an image of the entire sky. The name has been applied more recently to a class of extreme wide-angle lenses having enough barrel distortion to image a complete 180° in the object space on a finite circle of film. The law connecting the angular field in the object space with the size of the image is

$$h' = f'\theta \quad (\theta \text{ in radians}),$$

instead of the usual distortionless law $h' = f' \tan \theta$.

The earliest application of a fish-eye lens was in sky photography by meteorologists, who wished to record the entire cloud cover in a single picture. Such lenses were then called "sky lenses." Today fish-eye pictures are common in magazines and elsewhere when a full 180° of a scene is desired. The tilt of the entrance pupil of a fish-eye lens is discussed in Section II of Chapter 6.

V. SPECIAL PURPOSE LENSES

D. MONOCENTRIC BALL LENSES

A monocentric ball lens has no unique optical axis, and there is no limit to the extent of the angular field, the image lying on a spherical surface concentric with the center of the ball.

The first ball lens was the water-filled "Panoramic" lens constructed by T. Sutton in 1859. In 1944 James Baker designed an $f/3.5$ all-glass ball lens covering a 120° field with a resolution of over 200 lines/mm (Fig. 15.7). The image was recorded on a spherical glass photographic plate. A camera was constructed to use this lens, but it was too impractical for regular use. The illumination in this type of lens falls off by a factor equal to $\cos \theta$ instead of the usual $\cos^4 \theta$. It is possible that new uses for a lens of this type may arise in the future.

E. PHOTOGRAPHY IN WATER

Today a great deal of photography is performed under water, with the camera contained in a water-tight housing having a flat glass window in front. All types of aberration arise at this water–glass interface, the most obvious being distortion.

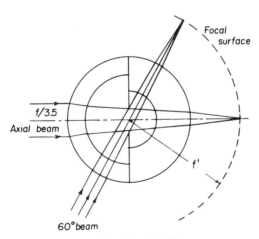

FIG. 15.7. Baker's ball lens.

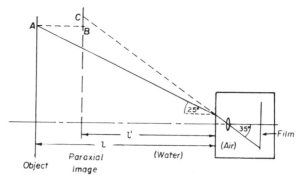

Fig. 15.8. In-water photography.

Figure 15.8 shows a typical situation in which the camera is capable of covering a field of ±35° in air but can cover only ±25° in water because the refractive index of water is about 1.33. If the object being photographed is at A, its paraxial image as seen by the camera would be at B, where $l/l' = 1.33$. The paraxial magnification at a plane refracting surface is unity, so the paraxial image of A would be at B. However, if we project the principal ray inside the housing out to intersect the image plane, it does so at C, which is higher than it should be, representing in this case a pincushion distortion of about 10%.

Many attempts have been made to correct these aberrations, the best procedure being to design a special lens such as that on the Nikonos camera. Other attempts have been directed toward the insertion of special correcting lenses between the plane front window and the camera.

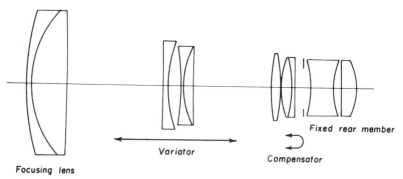

Fig. 15.9. Typical zoom lens for a small movie camera. (U.S. Patent 2,847,907)

VI. TYPES OF ZOOM LENSES

Zoom lenses of various types have become increasingly common during the past 30 years. They have also become increasingly complex, and their design would be a hopeless problem without the aid of computer optimization. Some of the chief types of zooms are discussed in this section.

A. MECHANICALLY COMPENSATED — WIDE RANGE

Mechanically compensated wide-range systems are used extensively on motion-picture and TV cameras. The focal length ranges from 3:1 to 30:1 (in extreme cases). Each system contains four basic components (Fig. 15.9):

(1) a positive front component, used to focus on near objects;
(2) a negative *variator* which moves through a considerable distance to vary the focal length of the system;
(3) a positive or negative *compensator*, moved by a cam to hold the image in a fixed focal plane;
(4) a fixed positive rear component carrying most of the power of the system. It also contains the iris diaphragm and often a beam splitter to reflect light into the viewfinder.

In some TV zoom lenses covering a wide focal-length range, the operation is divided into two parts, the first covering the shorter focal lengths at a fixed F-number and the second portion for the longer focal lengths at a diminishing relative aperture. Some TV zooms include the possibility of inserting a range changer behind the lens.

B. MECHANICALLY COMPENSATED — SHORT RANGE

Mechanically compensated short-range systems are exceptionally compact zoom lenses for use on an SLR camera. As the frame diagonal is about six times as large as the frame diagonal on a Super-8 movie camera, it is absurd to think of scaling up a movie zoom lens for use on

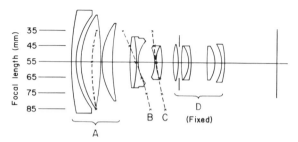

FIG. 15.10. Vivitar Series I zoom lens; 35 to 85 mm at $f/2.8$.

an SLR. Furthermore, the angular field of still camera lenses is much wider than that of a movie camera and the relative aperture is less(e.g., $f/3.5$ instead of $f/1.2$). There is no need to maintain the focal plane on a still camera because the user invariably adjusts the focus immediately before taking the picture anyway. For all these reasons the design of a still-camera zoom lens is very different from the design of a movie zoom, and it is not surprising to find that the focal-length range is approximately 2 : 1 instead of approximately 8 : 1 for the movie camera. Figure 15.10 shows a section of the recent Vivitar zoom lens for a 35-mm camera, where the focal-length range is from 35 to 85 mm at $f/2.8$.

C. Two-Component Zooms

Two-component zoom lenses are a recent development and are essentially a reversed telephoto with a variable spacing between the two components. They cover exceptionally wide angular fields, up to ±40° in some cases, although the type is often used for quite moderate fields. Both the front negative and rear positive components are moved during a zoom. The iris diaphragm is located in the rear component, and therefore the F-number changes with the focal length, but this is generally of no consequence with the modern use of automatic exposure control or matched pointers. Focusing on a near object is performed either by a movement of the whole system or by part of the front component. A typical objective of this type is shown in Fig. 15.11. This case was considered in Section III of Chapter 5.

VI. TYPES OF ZOOM LENSES

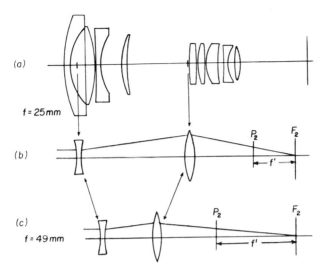

FIG. 15.11. A typical two-component zoom system.

D. Projection Zooms

Zoom lenses used on a projector may be of any type, and no attempt is made to hold the focus during a zoom. The focal length range is usually small, whereas the aperture is similar to that of a fixed lens intended for the same purpose.

E. Optically Compensated Zooms

Optically compensated zooms were an early development, based on the thought that if the cam in a mechanically compensated lens were to become worn, the focus might not remain in a fixed position. Generally these systems contained three components in order plus, minus, plus, with the two outer positive components coupled together. By careful design it was found possible to arrange the focal lengths and separations of the lenses so that the image remained reasonably fixed during a zoom. The image actually moved slightly in and out, but in small sizes this was insignificant. The focal length range was limited to 3:1 or 4:1 at most, and the type is now practically obsolete.

VII. SPECIFYING A PHOTOGRAPHIC OBJECTIVE

If a photographic lens is to be designed and constructed for some particular application, there are several important factors on which it is essential for the buyer and seller to agree, otherwise the lens may not meet the customer's requirements. It is particularly important that the purchaser should not overspecify, as every particular requirement that he may list renders the problem more difficult, the product more expensive, and lengthens the time required to design and make the lens. Every number that is stated must be accompanied by its tolerance, as nothing can be made to exact dimensions, and if the customer does not state his tolerances, someone will invent them for him, and they may not be at all what the customer requires.

Some of the factors to be considered are as follows:

(a) *Simple specification — distant object (or image).* The simple facts to be stated here are the desired focal length, angular field, and relative aperture (F-number).

(b) *Simple specification — finite magnification.* If a lens is to be used at finite magnification, it is of no significance to state the focal length or the angular field because these terms are meaningless. What must be stated is the distance from object to image, the sizes of object and image, and the effective F-number in either the object or the image space.

Following these simple specifications, other factors of importance are as follows:

(c) *Wavelength region.* The range of wavelengths to be under control must be known, and also the anticipated transmittance over this wavelength range.

(d) *Field coverage.* It is necessary to specify if the field is to be flat or curved, what the distortion tolerance is, and the permitted variation in illumination over the field (vignetting, $\cos^4$ law).

(e) *The number of lenses required and the anticipated cost per lens.* This requirement dictates whether low-cost glass, or even plastic materials, must be considered, the number of surfaces on a block, the possibility of using aspheric surfaces, and the lead time before lenses start being delivered to the customer.

(f) *Ambient conditions.* Temperature range, hostile environment, neutron flux, tropical atmosphere, underwater usage, protection against rain, vacuum operation, corrosive atmosphere, vibration tests, etc., are all ambient conditions that must be considered.

(g) *Image quality required.* This is likely to be the most difficult matter to be specified and agreed upon. Lens performance can be a matter of resolving power, MTF, encircled energy, image contrast and sharpness, etc., and each type of lens requires its own type of performance specification. The customer must never ask for a "perfect lens"; he must realize that this is impossible and expensive, and the greater margin he can accept in the way of definition, the easier it will be to meet his requirements.

(h) *Other tradeoffs.* Can a mirror be used, or must it be a lens? Could some kind of scanning system be used, or must the lens cover the whole field at once? Is a zoom lens required, or could a stepped power changer be used instead?

(i) *Mount details.* Should a focusing adjustment be provided? Is an iris diaphragm needed? Are there any mechanical interferences? Is the weight of the lens important?

(j) *Miscellaneous.* Under this heading come such matters as must the lens be symmetrical? Is it to be telecentric, or anamorphic? Is the image to be erect or inverted? Is it a visual, photographic, or projection system?

VIII. PANORAMIC OR SLIT CAMERAS

From the earliest days of photography, cameras have been constructed in which the image and the film move together at the same speed past a slit that acts as a focal-plane shutter. The principal types of slit camera are discussed in the following paragraphs.

A. THE SWINGING-LENS PANORAMIC CAMERA

The swinging-lens panoramic camera was invented by F. von Martens in 1845 for use with cylindrically curved daguerreotype plates. It was reintroduced in 1884 by P. Moëssard, using strips of flexible film. It was employed in the Kodak "Panoram" cameras dating from 1900 and in some modern cameras such as the Japanese "Panon." In this type of camera the lens is rotated during exposure about a vertical axis

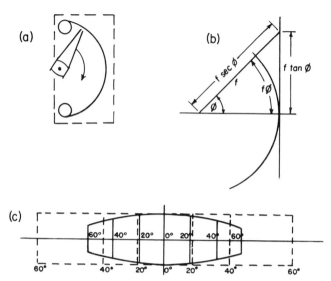

FIG. 15.12. The swinging-lens panoramic camera.

through the rear nodal point, the film being supported on a fixed cylindrical track centered about the vertical rotation axis of the lens. A tubular "funnel" joins the rear of the lens to a slit close to the film, the movement of the slit past the film acting as a focal-plane shutter (Fig. 15.12a). The exposure can be controlled by altering the rate of rotation of the lens about its axis, varying the F-number of the lens, or adjusting the width of the slit.

The definition in this type of camera is excellent because the lens has only to cover a field as wide as the vertical height of the slit, but the image is distorted in both the horizontal and vertical directions. The reason for this is indicated in Fig. 15.12b. The horizontal length of the finished picture is given by $f\phi$ instead of the usual $f\tan\phi$, where ϕ is the horizontal field angle, and the vertical dimension of the image is proportional to f instead of $f\sec\phi$. As a result, what should be a rectangle, as shown by dashed lines in Fig. 15.12c, is actually shortened and contracted at the ends, as shown by the solid lines.

B. The Rotating Camera

The rotating camera was originated by J. R. Johnson in 1862 under the name of the "Pantoscopic" camera. The whole apparatus rotates on its tripod about a vertical axis, and the film is driven past a slit in the focal plane at the same rate as the image is moving. The gearing to drive the film is chosen to suit the focal length of the lens and the distance of the object. A popular example of this type of camera was the Circut, dating from 1904.

C. The Aerial Slit Camera

In an aerial slit camera the lens axis points vertically downward toward the ground, and the image passes across a transverse slit in the focal plane as the airplane moves forward. The film is moved forward at such a rate that the image and film stay locked together during exposure. The film velocity depends, of course, on the speed of the airplane, the focal length of the lens, and the height of the plane above the ground, and in hilly country the velocity of the film must be continually adjusted to keep up with changes in the height of the terrain. This is performed automatically by electronic means that monitor the passage of ground details past a number of slits backed up by photocells.

Stereoscopic pictures can be made in this way by the use of two lenses mounted side by side in front of a single slit with one lens slightly ahead of the other. In this way the image of a detail on the ground is recorded on one side of the film before it is recorded on the other side and the distance that the airplane has moved between the two exposures constitutes the stereoscopic base. For viewing, one image is presented to the right eye and the other to the left.

D. The Race-Track Camera

A special form of slit camera is used to record the finish of a horse race. The camera is mounted high up above the track, looking obliquely downward, and a vertical slit is placed in the focal plane to record only events occurring along the finish line. The film is made to travel horizontally past the slit at about the speed of the image of the horses, although horses close to the camera will appear to move faster than those at the other side of the track. The important point is that any vertical line drawn across the completed film represents the finish line at a particular moment of time, hence a time scale can be drawn along the length of the film. The images of the horses will appear somewhat distorted because their legs will have moved between the times that their heads and their tails cross the finish line. A vertical mirror at the other side of the track permits the horses at that side to be photographed even though they are hidden behind those on the near side of the track.

E. Scanning Copy Cameras

In some microfilm cameras a moving document is photographed on a moving film traveling in the opposite direction with the ratio of the two movements equal to the magnification ratio of the lens (Fig. 15.13). A wide slit is inserted in the beam in front of either the document or the film to limit the extent of the document being photographed at any one time.

If an erecting lens is used at unit magnification between document and copy, it is possible to arrange matters so that the lens system moves between fixed document and copy planes, as in some machines for

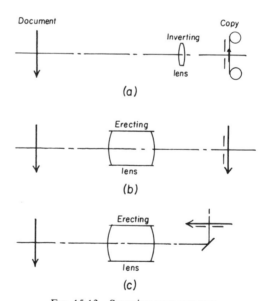

FIG. 15.13. Scanning copy cameras.

copying a mask on a silicon wafer in the integrated circuit industry. A variety of different imaging arrangements have been used in copying machines of the Xerox type.

F. ORTHOGRAPHIC PHOTOGRAPHY

It is occasionally necessary to make an *orthographic* photograph of some object, i.e., a photograph in which the beams of light are everywhere perpendicular to the mean plane through the object. Such a system must be telecentric in the object space.

If the field is small, it may be possible to find an existing telecentric lens that would meet the need. However, such lenses are rare, and it is generally easier to mount a large plano-convex field lens, often of plastic material, in front of the object and then place the camera at the rear focus of the field lens (Fig. 15.14).

If the object is larger than any available field lens, it may be necessary to use a scanning slit camera. The camera is moved along a horizontal track with a film moving at the correct speed behind a vertical slit in the image plane of the lens. After each scan the camera is moved down a small distance and another scan is made. If an erect-image lens is used, the whole film can be used as it comes out of the camera, but if an inverting lens is used, it is necessary to print each strip separately and to mount the strips in proper order to build up the complete scene.

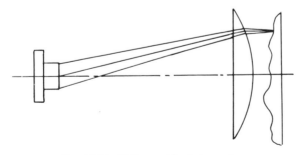

FIG. 15.14. Orthographic photography.

IX. MOTION-PICTURE SYSTEMS

In an ordinary motion-picture camera, a long strip of film, ranging from 8 to 70 mm in width, is mounted at the focus of a lens. One frame of film is exposed briefly to the image, and then a shutter obstructs the light while the film is moved rapidly to the next frame. The film is then held stationary during the next exposure and the cycle is repeated many times a second. Motion-picture film is perforated along one or both sides and each frame is represented by a fixed number of perforation holes. For example, in the standard 35-mm film used in theaters, there are four perforations to a frame, each frame being 16×22 mm in size. In 8- and 16-mm film there is only one hole per frame. The larger films are moved by an intermittent sprocket or some other suitable mechanism, whereas the smaller amateur films are generally moved by a simple claw or pusher device.

Ordinary intermittent motion-picture cameras are capable of frame rates up to perhaps 300 frames/sec, although the standard rates are 16 or 18 frames/sec for silent film or 24 for sound film. Movie projectors are generally limited to the standard frame rates.

The greatest care must be taken in all motion-picture equipment, cameras, printers, and projectors to ensure that the film remains absolutely stationary during exposure. The sprocket holes in the film and the various intermittent film movements must be equally precise to prevent any jiggle of the picture that would be disturbing to the audience.

A. High-Speed Cameras[1]

For frame rates higher than about 300/sec, the ordinary intermittent film movement becomes impractical, and various arrangements have been devised to cause the image to move continuously at the same speed as the film is moving in the camera, so that the next frame brightens up while the previous frame is fading out. The separation of the images along the film must, of course, be equal to the sprocket-hole spacing, which is 0.75 in. (19.05 mm) for 35-mm film, 0.3 in.

---

[1] A. Dubovik, "Photographic Recording of High-Speed Processes." Wiley, New York, 1981.

(7.62 mm) for 16-mm film, 0.167 in. (4.23 mm) for Super-8 film, and 0.15 in. (3.81 mm) for 8-mm film.

(*a*) *Rotating prism types.* The procedure generally adopted in a nonintermittent high-speed camera is to mount a rotating polygonal glass prism between the lens and the film. Each pair of opposite faces in such a polygon acts like a tilting parallel plate, and when the prism rotates in a clockwise direction, the image moves downward, and the film is made to follow the image at the same linear speed. The situation with an eight-sided prism is indicated in Fig. 15.15. In (a) the prism surfaces are shown perpendicular to the beam, and the image is centered on the axis at F_1. After the prism has rotated through 22.5°, the edges A and B come to the axis, and the beam from the lens is split into two equal parts, one-half forming the fading image F_1 while the other half forms the incoming image F_2. Soon F_1 disappears entirely and only F_2 is exposed. No shutter is needed in this case, although it is desirable to cut off the light at the extreme ends of each exposure because the image motion gradually speeds up and begins to be faster than the film, causing blurring of the picture.

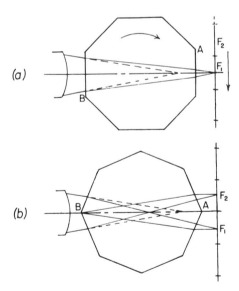

FIG. 15.15. A refracting polygon in a movie camera.

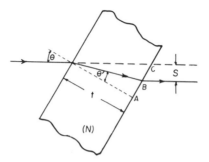

FIG. 15.16. Geometry of a ray passing through a thick glass plate.

The reason for this is seen in Fig. 15.16, where a tilted parallel plate is indicated behind a lens. If the tilt angle is θ and the refractive index is N, then the image shift S is given by

$$S = t\left[\frac{\sin(\theta - \theta')}{\cos \theta'}\right] = t(\tan \theta - \tan \theta')\cos \theta$$

$$= t \sin \theta\left[1 - \frac{1}{N}\frac{\cos \theta}{\cos \theta'}\right].$$

For a very small tilt of the plate, this expression reduces to its paraxial form, and

$$s = t\left[1 - \frac{1}{N}\right],$$

which is less than the accurate value S. The gradual increase in S as the tilt is increased is shown graphically in Fig. 15.17, which refers to a

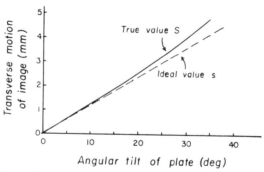

FIG. 15.17. Values of S and s for a 20-mm-thick plate of index 1.523.

plate 20 mm thick having a refractive index equal to 1.523. Evidently, in such a case, the plate should not be allowed to turn through an angle greater than approximately 15°, or blurring starts to occur. Another problem, not shown in these diagrams, is that the image gradually becomes distorted in shape at increasing tilt angles, which is another reason for using a shutter to cut off the light between frames.

The design of a camera or projector using a polygonal prism requires some care. The faces of the prism must be large enough to cover the whole frame on the film, and the thickness must be such that s will be equal to the frame height when θ is equal to half the angle between facets. These conditions enable the refractive index to be determined for any given number of facets on the prism. The more facets that are used, the better the image quality will be, and the slower must be the angular speed of the prism. But a large prism requires a long space between the back of the lens and the film, so either a lens having a long back focus (such as the lens from an SLR camera) must be used or a relay must be employed to reimage the first image onto the film. Both

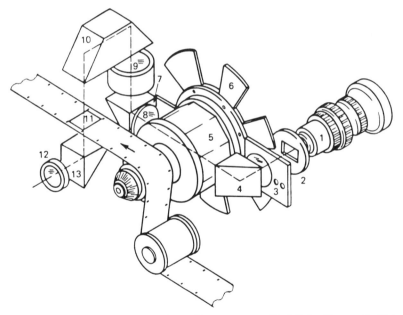

FIG. 15.18. Optical system of the Hycam camera (Red Lake Corporation): 1. objective lens; 2. aperture plate; 3. first field lens; 5. compensating polygon; 6. shutter; 8. second field lens; 9. relay lens; 11. image on film; 12–13. viewfinder eyepiece.

these procedures have been used in commercial high-speed cameras. Frame rates up to 10,000/sec have been achieved on 16-mm film, the limit being reached when it is no longer possible to unwind and rewind the film fast enough without damage to the film. Often a large part of the film in a camera is used up in merely reaching the maximum speed, before attempting to record the event which is to be photographed.

Originally the rotating prism and the film drive mechanism were on separate shafts connected by gearing, but recently it has been found possible to mount both drives, and also a shutter, on the same shaft by inserting fixed prisms in the optical path, as shown in Fig. 15.18.

If fewer facets are used to make a smaller prism requiring a smaller back focal clearance, it is necessary to include shutters between successive prism faces to cut off the light as soon as the image quality begins to deteriorate. Such an arrangement for a square prism is indicated in Fig. 15.19.

(b) *Rotating mirror types.* For faster frame rates, a short length of film is mounted inside a drum and a rapidly rotating three- or six-sided

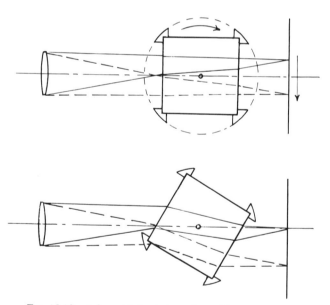

FIG. 15.19. A four-sided rotating prism with cutoff shields.

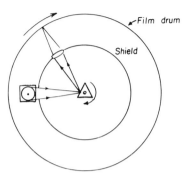

FIG. 15.20. Rotating-mirror camera.

polygonal mirror is located at the center of the drum (Fig. 15.20). An image of the event is formed on the rotating mirror so that a beam of light sweeps around inside the drum as the mirror rotates. Everything is shielded from the light except for a relay lens which forms an image on the film every time the beam sweeps past it. By rotating the drum at a suitable speed, the successive images will be spaced apart along the film.

In some cameras two or more relay lenses are used, each lens being slightly displaced at right angles to the film so as to form another row of images alternating with those in the first row. As many as four rows of images have been used on a single strip of wide film. Frame rates up to 35,000/sec have been achieved with a single relay lens, and more when several relays are used.

For even higher frame rates the film is kept stationary, and a row of closely spaced relay lenses are used, the sweeping beam of light entering each relay lens in turn to form a row of images along the film (Fig. 15.21). This arrangement is known as *aperture shuttering.* Frame rates up to 25 million/sec have been achieved, but of course only a few frames can be recorded along a limited length of film, and synchronizing the event with the camera becomes a major problem.

(c) *Streak cameras.* For some purposes it is desired to record merely the size and duration of an event, such as the explosion of an atomic bomb, and then a streak camera is useful. An image of the event is formed on a narrow slit, which in turn is imaged on a strip of film inside a rapidly rotating drum, the long dimension of the slit lying

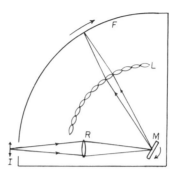

Fig. 15.21. Ultra-high-speed camera with aperture shuttering.

across the film. Some high-speed cameras have been constructed in which a streak record and a framing record are formed simultaneously on two separate strips of film. In any camera containing only a short length of film, a capping shutter must be added, which is opened immediately before the event and closed immediately after it to prevent the accidental formation of overlapping exposures.

B. A Flying Spot Scanner

The flying spot scanner is an arrangement used in television to transmit a slide transparency or a frame of motion pictures. The simplest procedure is to generate a TV raster on the face of a cathode-ray tube and place the transparency close to the tube; a photocell located at a reasonable distance picks up the light transmitted by the slide as the cathode-ray spot traverses the raster.

However, because of the thickness of the glass at the face of the tube, the transparency is necessarily slightly out of focus, and therefore the image is not quite sharp. A better procedure is to use a two-lens Köhler system (Fig. 15.22). Here the raster R is imaged accurately on the slide

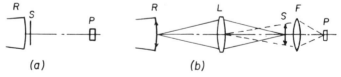

Fig. 15.22. Two forms of flying spot scanner.

S by means of the lens L, and a field lens F behind the slide forms an image of the lens aperture on the photocell at P. In this way the lens is filled with light at all times, no matter where the cathode-ray spot may be, and its image fills the photocell at all times.

CHAPTER 16

Spectroscopic Apparatus

I. DISPERSING PRISMS

Dispersing prisms have been used in spectroscopic apparatus for over a century and are still used in preference to gratings because of their ease of manufacture if only a single spectral order is required and if the prism material is transparent to the range of wavelengths under investigation. The necessary formulas for determining the dispersion and resolving power of a prism follow.[1]

A. Prism Deviation

It is clear from Fig. 16.1, which shows a single ray passing through a prism, that

$$\sin \beta = (\sin \alpha)/n,$$
$$\gamma = A - \beta, \tag{1}$$
$$\sin \delta = n \sin \gamma.$$

The ray deviation is

$$D = \alpha + \delta - A, \tag{2}$$

where A is the prism angle, n its refractive index, and α, β, γ, and δ the successive values of the angles of incidence and refraction at the two surfaces of the prism.

If we are given the prism angle A and the angle of incidence α, we can determine the angle of emergence δ at the second surface by

$$\sin \delta = n \sin(A - \beta) = \sin A \sqrt{n^2 - \sin^2 \alpha} - \cos A \sin \alpha. \tag{3}$$

A plot of the relation between deviation and the angle of incidence in a typical prism is shown in Fig. 16.2.

[1] R. Kingslake, Dispersing prisms, *in* "Applied Optics and Optical Engineering" (R. Kingslake, ed.), Vol. 5, p. 12. Academic Press, New York, 1969.

I. DISPERSING PRISMS

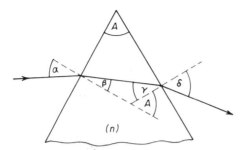

FIG. 16.1. A refracting prism.

Minimum Deviation. The graph in Fig. 16.2 shows that at some particular angle of incidence the ray deviation reaches a minimum. To determine this value, we differentiate Eq. (2) with respect to α giving

$$\frac{\partial D}{\partial \alpha} = 1 + \frac{\partial \delta}{\partial \alpha},$$

and at minimum deviation, obviously $\partial \delta / \partial \alpha = -1$. The value of $\partial \delta / \partial \alpha$ can be found from the three expressions in Eq. (1) by

$$\frac{\partial \delta}{\partial \alpha} = \frac{\partial \delta}{\partial \gamma}\frac{\partial \gamma}{\partial \beta}\frac{\partial \beta}{\partial \alpha} = \frac{n \cos \gamma}{\cos \delta}(-1)\frac{\cos \alpha}{n \cos \beta} = -\frac{\cos \alpha \cos \gamma}{\cos \beta \cos \delta}.$$

Hence at minimum deviation,

$$\cos \alpha \cos \gamma = \cos \beta \cos \delta.$$

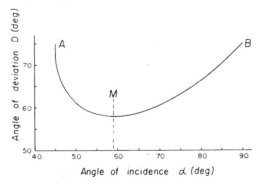

FIG. 16.2. Relation between angle of incidence and angle of deviation for a 60° prism, with $n = 1.7159$.

Converting the cosines into sines by the square root and using the fundamental relationships in Eq. (1), we find that at minimum deviation,

$$\alpha = \delta \quad \text{and} \quad \beta = \gamma.$$

That is, the ray passes symmetrically through the prism and crosses the bisector perpendicularly.

Also at minimum deviation,

$$\alpha = \tfrac{1}{2}(D + A) \quad \text{and} \quad \beta = \tfrac{1}{2}A,$$

from which

$$n = \sin\alpha/\sin\beta = \sin\tfrac{1}{2}(D + A)/\sin\tfrac{1}{2}A.$$

This relation is commonly used when measuring the refractive index of a prism on a laboratory spectrometer.

B. Prism Dispersion

By the *dispersion* of a prism is meant the change in the emerging ray slope caused by a change in refractive index from one wavelength to another. Expressed more exactly, we say that

$$\text{dispersion} = \frac{\partial \delta}{\partial \lambda} = \frac{\partial \delta}{\partial n}\frac{\partial n}{\partial \lambda}.$$

Differentiating Eq. (2) with respect to n, regarding the prism angle A and the angle of incidence α as constant, we find that, by Eq. (3):

$$\cos\delta \, \frac{\partial \delta}{\partial n} = \frac{\tfrac{1}{2}\sin A}{\sqrt{n^2 - \sin^2\alpha}} \cdot 2n = \frac{\sin A}{\cos\beta},$$

from which

$$\frac{\partial \delta}{\partial n} = \frac{\sin A}{\cos\beta \cos\delta},$$

and the prism dispersion is

$$\text{dispersion} = \frac{\partial \delta}{\partial \lambda} = \frac{\sin A}{\cos\beta \cos\delta}\frac{\partial n}{\partial \lambda}. \tag{4}$$

To illustrate the use of this formula, we consider the case of a 60° prism made of Schott's dense flint glass SF-1, set at minimum deviation for the wavelength 0.60 μm. Using the interpolation formula given in Schott's catalog, we can determine the refractive index of this glass for a series of wavelengths, and by differentiating the formula we can also calculate the successive values of $\partial n/\partial \lambda$ as shown in the following tabulation.

λ (μm)	0.4	0.5	0.6	0.7	0.8	0.9	1.0
Refractive index n	1.76427	1.73152	1.71589	1.70692	1.70115	1.69712	1.69409
$\partial n/\partial \lambda$	−0.50159	−0.21274	−0.11433	−0.07021	−0.04740	−0.03450	−0.02676

As the prism is set at minimum deviation for the wavelength 0.60 μm, we find from the above formulas that the fixed value of α is 59.0866°, and this is also the value of δ at $\lambda = 0.60$ μm.

For substitution in Eq. (4), we see that $\sin A/\cos \beta = 1.0$, and hence for this case,

$$\text{dispersion} = \frac{\partial n}{\partial \lambda} \sec \delta.$$

The value of δ at each wavelength has to be determined by Eq. (3), giving the values in the following tabulation.

δ (deg)	64.9726	60.8740	59.0866	58.0995	57.4778	57.0494	56.7303
$\sec \delta$	2.3638	2.0545	1.9465	1.8923	1.8600	1.8385	1.8229
dispersion	−1.1856	−0.4371	−0.2225	−0.1329	−0.0882	−0.0634	−0.0488

Dispersion is equal to the product of the values of $\sec \delta$ and the values of $\partial n/\partial \lambda$. The units for the last row of values are radians/μm. They can, of course, be converted into any other units if desired.

C. Prism Magnification

We saw in Section III of Chapter 3 that the width of a parallel beam refracted at a single plane surface is given by

$$\frac{D}{d} = \frac{\cos \alpha}{\cos \beta},$$

and hence, after the two faces of a refracting prism, the width of the parallel beam is given by

$$\frac{D}{d} = \frac{\cos \alpha \cos \gamma}{\cos \beta \cos \delta}.$$

This tells us that at minimum deviation the prism magnification is unity, and if we place the prism in front of our eye, the image that we see will have the same shape as the object, although it will be deviated to one side and highly colored because of the deviation and dispersion of the prism.

D. Resolving Power of a Prism

Assuming a diffraction-limited optical system, an infinitely narrow slit, and perfectly parallel light passing through the prism, it is customary to define the resolving power (RP) of the prism as

$$\text{RP} = \lambda/d\lambda,$$

where λ is the wavelength of the light and $d\lambda$ the wavelength separation between two adjacent spectral lines that are just resolved by the instrument.

The situation is illustrated in Fig. 16.3. We assume that the width of the refracted beam is limited by the objective lens and that the prism is larger than necessary to accommodate the whole beam.

By diffraction theory, the angular resolution of the objective lens is λ/w radians, hence

$$\lambda/w = d\delta = \frac{\partial \delta}{\partial \lambda} \cdot d\lambda,$$

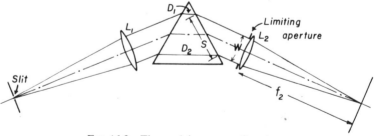

Fig. 16.3. The resolving power of a prism.

where $d\delta$ is the angular separation of the just resolved spectral lines having wavelengths λ and $\lambda + d\lambda$. Hence the resolving power is given by

$$\text{RP} = \frac{\lambda}{d\lambda} = w\frac{\partial \delta}{\partial \lambda}.$$

The RP of the prism is therefore equal to the product of the dispersion of the prism and the width w of the emerging beam.

By Eq. (4)

$$\frac{\partial \delta}{\partial \lambda} = \frac{\sin A}{\cos \beta \cos \delta}\left(\frac{\partial n}{\partial \lambda}\right),$$

and as $w/\cos \delta$ is the slant length s of the beam on the prism face, we see that

$$\frac{\lambda}{d\lambda} = s\frac{\sin A}{\cos \beta}\left(\frac{\partial n}{\partial \lambda}\right).$$

If the prism is in minimum deviation, which is the usual situation, $\beta = \frac{1}{2}A$ and the RP becomes

$$\frac{\lambda}{d\lambda} = 2s \sin \frac{1}{2}A \left(\frac{\partial n}{\partial \lambda}\right).$$

$2s \sin \frac{1}{2}A$ is the difference between the longest and shortest paths within the prism, namely, $D_2 - D_1$. Therefore the RP is given by

$$\frac{\lambda}{d\lambda} = (D_2 - D_1)\frac{\partial n}{\partial \lambda}$$

As an example, to resolve the two sodium lines requires an RP of 980. If the prism is made of Schott's SF-1 glass, the value of $\partial n/\partial \lambda$ at the sodium wavelength is about 0.12; hence the difference $D_2 - D_1$ must be not less than 8170 μm, or, say, 9 mm. A prism of 60° angle and ~ 0.5 in. on a side made of this glass should easily resolve the sodium lines.

E. Curvature of Spectrum Lines

So far we have considered only events occurring in the principal section of a prism perpendicular to the refracting edge. However, in a spectroscopic instrument equipped with a slit of finite length, light from the two ends of the slit traverses the prism obliquely. For such an oblique beam, the height of the prism is the same as in the principal section but the base length is slightly increased, so the prism angle is effectively a little greater for the ends of the slit than for the middle. This causes the spectrum lines to be curved with the ends displaced toward the blue end of the spectrum.

A formal analysis of this situation has been given in the literature,[1] and it is shown there that the radius of curvature of a spectrum line is given by

$$\rho = f' \frac{n}{n^2 - 1} \left(\frac{\cos \beta \cos \delta}{\sin A} \right),$$

where f' is the focal length of the objective lens. The focal length of the collimator lens is immaterial. For a 60° prism in minimum deviation, this reduces to

$$\rho = f' \frac{n}{n^2 - 1} \sqrt{1 - \frac{n^2}{4}}.$$

Thus, for the SF-1 glass that we have been considering and with an objective lens of 10-in. focal length, the radii of curvature of the spectral lines are given in the following tabulation.

λ (μm)	0.4	0.5	0.6	0.7	0.8	0.9	1.0
n	1.7643	1.7315	1.7159	1.7069	1.7012	1.6971	1.6941
ρ (in.)	3.933	4.337	4.534	4.649	4.723	4.776	4.815

We see that the curvature increases steadily from the red to the blue end of the spectrum.

F. THE PRISM SPECTROGRAPH[2]

The simplest form of prism spectrograph consists of a slit at the focus of a collimator lens to give a parallel beam, a prism, and an

[2] R. J. Meltzer, Spectrographs and monochromators, *in* "Applied Optics and Optical Engineering" (R. Kingslake, ed.), Vol. 5, p. 47. Academic Press, New York, 1969.

I. DISPERSING PRISMS

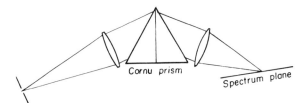

FIG. 16.4. A typical quartz spectrograph.

objective lens to form a spectrum. The spectrum can be recorded on a photographic film, or selected wavelengths can be detected by photocells placed appropriately along the spectrum.

In spite of the fact that the dispersion can be increased by tilting the prism out of minimum deviation, this is seldom done because the image magnification is then greater than unity, so the resolution is not increased although the dispersion is, and because it is only at minimum deviation that the location of the spectrum is independent of slight errors in mounting the prism in the instrument.

If the collimator and objective lenses are achromatic, the spectrum is perpendicular to the lens axis, although, because of the presence of secondary spectrum, the ends of the spectrum will be bent slightly backward, particularly at the blue end. For work in the ultraviolet, a Cornu prism with simple quartz lenses is often used, and in that case the blue end of the spectrum is formed much closer to the objective lens than the red end and the whole spectrum is steeply tilted relative to the lens axis (Fig. 16.4).

To reduce the size of the instrument and to save one lens, it is possible to use a 30° prism aluminized on the rear surface to autocollimate the light. To prevent the spectrum from being obstructed by the

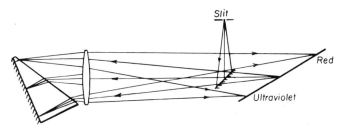

FIG. 16.5. A quartz Littrow spectrograph.

incident light, the slit is raised slightly above the plane of the emerging light, and a small mirror is inserted near the slit to bend the beam through a right angle, as shown in Fig. 16.5. Such an arrangement is known as a Littrow spectrograph. The objection to this arrangement is that stray light caused by reflection from the lens and prism surfaces often falls on the film, and if the light is of a wavelength to which the film is particularly sensitive, it may give a false impression of the spectrum being examined. Coating the surfaces and slightly tilting the lens can help matters somewhat.

G. A Prism Monochromator

A prism spectrograph can be converted into a source of monochromatic light by placing an exit slit at the desired wavelength in the plane of the spectrum. However, it is generally desired to vary the wavelength of the emitted light without moving either slit, and in that case the Littrow arrangement is particularly convenient because it is merely necessary to rotate the aluminized half-prism to cause the spectrum to traverse the exit slit and thus vary the emitted wavelength. Either a line source or a continuous white-light source can be used. If a line source is used, the spectral lines will be curved, and a curved exit slit is required. Because the curvature varies with wavelength, the exit slit should be made to fit a line in about the middle of the spectrum, with the hope that it will not be too far off at the two ends.

If for any reason the Littrow arrangement is undesirable, the Wadsworth system may be used. In this arrangement the 60° prism is mounted on a turntable with a plane mirror, the axis of rotation being at A, the intersection of the mirror and the bisector of the prism, as indicated in Fig. 16.6. Rotating the turntable causes the spectrum to traverse the exit slit in such a way that whatever wavelength is emitted, that is the wavelength which has passed through the prism in minimum deviation, so the width of the slit image remains constant throughout the spectrum. Lenses can be used for the collimator and objective, but for use over a wide spectral range it is better to use mirrors, because the inevitable secondary spectrum of a lens would cause most of the wavelengths to be slightly out of focus. The mirrors are off-axis paraboloids, and the instrument can be used in the UV or IR spectral regions if desired.

I. DISPERSING PRISMS

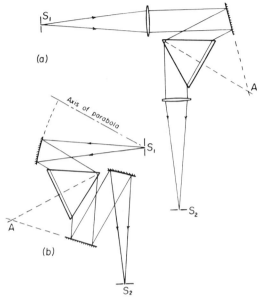

FIG. 16.6. Wadsworth monochromators with (a) lenses and (b) mirrors.

H. THE PELLIN–BROCA PRISM

The Pellin–Broca prism is a modification of the Wadsworth arrangement in which the 60° prism is divided into two halves with a plane mirror at 45° between them. The prism can be composite, as shown in Fig. 16.7a, or it can be made from a single piece of glass. Any ray entering at such an angle α that its angle of refraction β is 30° crosses the internal interfaces perpendicularly and is thus equivalent to a ray through a 60° prism at minimum deviation. The portions of this ray inside the prism are, respectively, parallel and perpendicular to the end face AB. This internal ray establishes the point of emergence, and of course the emergent ray is perpendicular to the entering ray.

To change the emitted wavelength it is only necessary to rotate the prism. However, there is one point of rotation for which the entering and emerging rays do not move laterally; it is at the point where the mirror face intersects the bisector of the angle between the entering and emerging portions of the principal ray, at O (Fig. 16.7b). The pivot

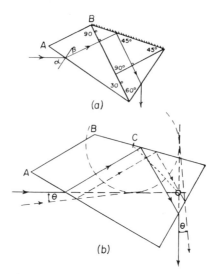

FIG. 16.7. The Pellin–Broca prism: (a) made from three pieces of glass; (b) rays in two different wavelengths.

point C in this diagram is the axis of rotation for this example. It is seen that the circle centered at C is tangent to the entering and emerging portions of both the solid and dashed rays in the diagram, which represent two different wavelengths in the beam. These rays are parallel internally and strike the mirror surface at different points.

One great convenience of the Pellin–Broca prism is that for the minimum deviation case the entering and emerging portions of the ray are at right angles. Hence the collimator and telescope of a monochromator using this prism must be fixed at right angles to each other. Thus, whatever light emerges from the exit slit has passed through the equivalent 60° prism at minimum deviation, with all the advantages of this situation. For this reason it has been called a "constant deviation prism," but this is misleading; the whole instrument should be called a "constant deviation spectroscope."

I. A Double Monochromator

In some applications the slightest trace of stray light of a different wavelength mixed with the useful light is extremely undesirable. For instance, in an experiment to determine the limit of sensitivity of a photographic emulsion in the infrared, the slightest trace of blue light superposed on the desired red light would completely vitiate the measurements.

If the stray light in a monochromator, from any cause, amounts to, say, 4% of the desired light at any setting of the wavelength scale, then using two monochromators in succession would reduce the stray light to 4% of 4%, or only 0.16%, while scarcely affecting the intensity of the desired wavelength. A double monochromator generally consists of two identical monochromators in tandem, sharing a common central slit. The two instruments may be additive or subtractive, depending on whether the second instrument increases or opposes the dispersion of the first instrument and increases or reduces the curvature of the spectral lines. The subtractive instrument is in many ways more convenient. Consider, for example, the van Cittert double monochromator shown in Fig. 16.8. Here the first monochromator forms a spectrum at S_2, which is recombined in the second instrument to form white light at the straight exit slit. A shaped mask, or a series of slits, mounted in the plane of the middle spectrum isolates some wavelengths, which are then combined to give the resulting colored light at the exit slit. To convert the instrument into a normal double monochromator, it is merely necessary to insert a suitably curved slit into this spectrum, and move it along the spectrum to isolate any desired wavelength at the exit slit. It is necessary to insert a pair of field lenses near the spectrum to form an image of one prism on the other to eliminate vignetting.

Any double monochromator eliminates unwanted wavelengths from the emerging light because the second instrument emits only light which falls into its proper place in the middle spectrum. Thus unwanted blue light mixed with the red light at the middle spectrum,

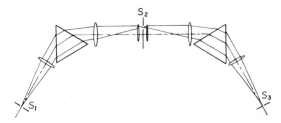

FIG. 16.8. The van Cittert subtractive double monochromator.

for example, is imaged somewhere on the jaws of the exit slit, and so cannot emerge. Two Wadsworth monochromators have been combined to form a double monochromator by mounting the two prisms and their associated plane mirrors on two turntables mounted one above the other on a common rotation axis, the curved central slit being placed horizontally and fed with light by two small mirrors at 45°, one under and one over the central slit.

This is a good place to remark that the use of several mirrors in a system may lead to considerable loss of light at wavelengths for which the reflectivity of the coating material is low. This remark does not apply to the infrared, because in that spectral region the reflectivity of aluminum is high at all wavelengths. There are no good reflectors in the ultraviolet.

II. DIFFRACTION GRATINGS[3]

The simplest concept of a plane diffraction grating is a row of equidistant slits transmitting a plane wave of light parallel to the plane containing the slits. Each slit emits a series of Huygenian wavelets which expand in all directions. The diffracted plane wave fronts are envelopes of these wavelets, as indicated in Fig. 16.9. The directions of the successive plane waves will be given by

$$\sin \theta = k\lambda/s, \qquad (5)$$

where λ is the wavelength of the light (the radius of the first wavelet in each case) and s the separation between the slits. The factor k is an integer and the value of k is the "order" of the diffraction. Because the value of θ varies with wavelength, a spectrum is formed in each order when white light is incident upon the grating.

As an example, Table I relates to the case where there are 300 slits/mm, and white light ranging from 0.4 μm in the blue to 0.7 μm in the red is incident on the grating. The values of θ for each wavelength in each order are tabulated, using the formula given in Eq. (5). The results from Table I are indicated pictorially in Fig. 16.10. It is

---

[3] D. Richardson, Diffraction gratings, *in* "Applied Optics and Optical Engineering" (R. Kingslake, ed.), Vol. 5, p. 17. Academic Press, New York, 1969; G. W. Stroke, Diffraction gratings, *in* "Handbuch der Physik" (S. Flugge, ed.), Vol. 29, p. 426 (in English). Springer Verlag, Berlin and New York, 1967.

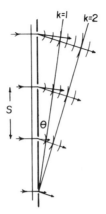

FIG. 16.9. The first two spectral orders in one wavelength.

immediately evident that there will be some degree of overlapping between the successive orders, particularly when k is large, only the first order shown here being free from overlapping. For this reason, the first order is used in most applications, although there is greater dispersion in the higher orders.

TABLE I

ANGULAR SPREAD OF SUCCESSIVE SPECTRAL ORDERS

λ	θ (deg)			
	$k=1$	$k=2$	$k=3$	$k=4$
0.0004	6.9	13.9	21.1	28.7
0.0005	8.6	17.4	26.7	36.9
0.0006	10.4	21.1	32.7	46.1
0.0007	12.1	24.8	39.0	57.1

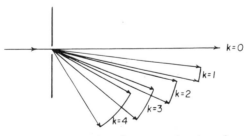

FIG. 16.10. The extent of the first four spectral orders, from $\lambda = 0.0004$ to 0.0007 mm, with $s = 0.00333$ mm.

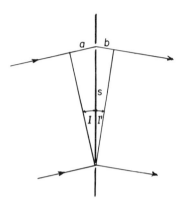

FIG. 16.11. Incident light at an angle to a grating.

If the incident plane light wave is inclined to the grating by an angle of incidence I, the angle of diffraction I' is related to the angle of incidence by

$$s(\sin I + \sin I') = a + b = k\lambda, \qquad (6)$$

where a and b are the distances of the entering and emerging wave fronts from the first slit. The situation is indicated in Fig. 16.11.

So far, the discussion has been restricted to a transmission grating, but the same arguments can be applied to a reflecting grating ruled on a layer of aluminum evaporated on a glass substrate. Such a grating acts like a mirror so far as the zero order is concerned, the diffracted orders lying on both sides of the reflected zero order, with the blue ends of the spectra closest to the zero order and the red ends further out; the second and higher orders, of course, fall beyond the first order spectra, as usual.

A. Image Magnification

As with a prism, a diffraction grating can form images that are wider or narrower than the original slit, the image magnification being given by

$$m = \frac{\partial I'}{\partial I} = -\frac{\cos I}{\cos I'},$$

II. DIFFRACTION GRATINGS

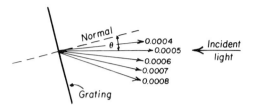

FIG. 16.12. Reflected spectrum by autocollimation.

a result found by differentiating Eq. (6). As with a prism, a grating also forms slit images that are curved, for the same reason.

Evidently, there will be no slit-image magnification if $I = I'$, that is, if the entering ray and the diffracted rays are coincident. This is known as the autocollimated arrangement, the grating being tilted at an angle θ to the entering light given by

$$\sin \theta = \frac{k\lambda}{2s}.$$

This applies, of course, to only one wavelength, usually in the middle of the spectrum. For other wavelengths we have, by Eq. (6),

$$\sin I' = -\sin I \pm \frac{k\lambda}{s}. \tag{7}$$

For the zero order, this becomes $I' = -I$, which is the ordinary law of reflection. The various diffracted images are spaced away from the reflected ray by the factor $k\lambda/s$ in Eq. (7).

As an example, suppose that $s = 0.001$, i.e., 1000 grooves/mm, and suppose that the grating is in the autocollimated position for $\lambda = 0.0005$ mm (see Fig. 16.12). Then the grating tilt is 14.48°, and the other wavelengths have the values of angle I' shown in the following tabulation.

λ (mm)	0.0004	0.0005	0.0006	0.0007	0.0008
I' (deg)	8.63	14.48	20.49	26.74	33.37

Of course, there will be another diffracted spectrum at equal slope on the other side of the mirror normal, but this is generally not used.

B. Dispersion

The angular dispersion of a grating is found by differentiating Eq. (7) with respect to wavelength. Because I is constant, we have

$$\text{dispersion} = \partial I'/\partial \lambda = k/s \cos I'.$$

C. Resolving Power

The resolving power of a grating, as for a prism, is given by $\text{RP} = \lambda/d\lambda$, where $d\lambda$ is the just resolved difference between two adjacent spectral lines at a wavelength equal to λ. To find its magnitude, we refer to Rayleigh's rule that two adjacent line images will just be separated if one image falls on the first dark ring of the other. The angular resolution is then equal to λ/A rad, where A is the diameter of the lens aperture. Referring to Fig. 16.13, we see that for a grating

$$A = Ns \cos I',$$

where N is the number of lines in the grating and s the line separation. Equating Rayleigh's resolution to the dispersion, we obtain

$$\frac{\lambda}{Ns \cos I'} = \frac{k d\lambda}{s \cos I'},$$

from which the RP becomes

$$\lambda/d\lambda = Nk,$$

which is the simple product of the number of grooves in the grating and the order of the spectrum. If W is the width of the grating, $W = Ns$, and since $k = s(\sin I + \sin I')/\lambda$ by Eq. (7), we see that the RP becomes

$$\frac{\lambda}{d\lambda} = \frac{Ns(\sin I + \sin I')}{\lambda} = \frac{W}{\lambda}(\sin I + \sin I').$$

As the maximum possible value of $\sin I + \sin I'$ is 2.0, it is clear that the maximum possible RP of a 6-in. grating at $\lambda = 0.0005$ mm is 600,000, regardless of the spectral order or the number of grooves in the grating.

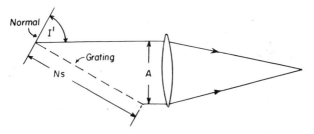

FIG. 16.13. Angular resolution of a plane grating.

D. Blaze

By suitably shaping the grooves in a grating, it is possible to direct most if not all of the light into one specific direction, which, of course, is made to coincide with the desired diffraction image. Thus, for example, in the autocollimated case, the grooves should be perpendicular to the incident light, hence they should be flat and inclined to the face of the grating by an angle $\theta = I$ (see Fig. 16.14). Such a grating is said to be *blazed* for this particular application. A transmission grating can also be blazed, but the blaze angle must be much larger.

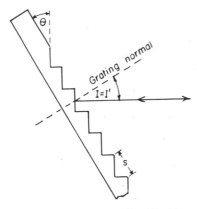

FIG. 16.14. A blazed reflecting grating. The blaze angle is θ.

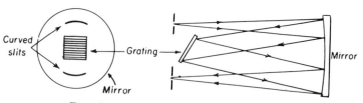

Fig. 16.15. The Ebert–Fastie monochromator.

E. A Plane-Grating Spectrograph

A typical spectrograph equipped with a plane reflecting grating is the Ebert–Fastie arrangement shown in Fig. 16.15. The large concave mirror serves as the collimator and image former, with the grating set at a suitable angle in between. The grating is thus nearly in the autocollimated position, and the slit-image magnification is very close to unity. The instrument can be converted into a monochromator by installing an exit slit at the middle of the spectrum; rotating the grating will then cause the spectrum to traverse the exit slit and let any desired wavelength pass through it. Both slits should be curved about the middle axis of the mirror, as indicated in Fig. 16.15. Several other possible applications of a plane grating are described in the literature.

F. A Concave Grating

It occurred to H. A. Rowland that by ruling the grating on a concave substrate the focusing action and the dispersion could be combined in a single unit (Fig. 16.16). It is easily shown that the slit and the spectral images in all wavelengths fall on a circle called the Rowland circle; the center of curvature of the grating also lies on this circle.

In Fig. 16.16 the reflected ray from the mirror is shown; this constitutes the zero-order spectrum. The effect of the grating is to generate two series of spectra, first-order, second-order, etc., on both sides of the reflected ray (only one first-order spectrum is shown in this diagram). The grating blaze would, of course, be adjusted to send as much light as possible into this order.

In this illustration, the angle of incidence is 20° and there are assumed to be 667 grooves/mm on the grating, i.e., $s = 0.0015$ mm. Then the angles of diffraction, measured from the normal, are given in

II. DIFFRACTION GRATINGS

TABLE II
Dispersion of a Concave Grating

Spectrum	λ (mm)	I' (deg)
Blue	0.0004	−4.3
	0.0005	−0.5
	0.0006	+3.3
	0.0007	+7.2
Red	0.0008	+11.0

Eq. (7), namely,

$$\sin I' = -\sin I \pm \frac{k\lambda}{s}.$$

Only the plus sign is shown in this diagram, and of course, $k = 1$ for the first order. A series of values of I' are given in Table II. When I' has a negative sign it means that the diffracted ray is on the same side of the normal as the zero order, while a positive angle I' implies that the diffracted ray is approaching the incident light as shown in Fig. 16.16. The second order would pass beyond the incident light. A number of different forms of mounting for concave gratings have been described, each of which has its virtues and its disadvantages. At the present time, however, it is generally found preferable to use a plane grating with some lens or mirror arrangement to focus the light into a spectrum or a monochromator as required.

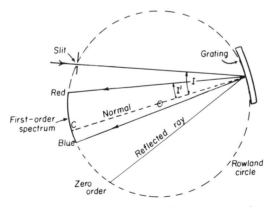

FIG. 16.16. The spectrum formed by a concave grating.

G. Acousto-Optic Devices

It was shown by Debye and Sears, in about 1932, that a supersonic wave train traveling through a liquid can be used as a phase grating, to deviate or disperse light. The arrangement is illustrated diagrammatically in Fig. 16.17. If the liquid layer is thin, it acts like an ordinary plane diffraction grating, but if it is thick, the situation becomes very complicated and the liquid layer acts more like x-ray Bragg diffraction in a crystal. For a thin liquid layer the angle of diffraction θ is given by $\sin \theta = k\lambda/\Lambda$, as in an ordinary grating, where k is an integer, λ the wavelength of the light, and Λ the wavelength of the sound in the liquid. The fact that the sound waves are moving rapidly in a transverse direction introduces a Doppler shift into the acoustic grating. The higher the intensity of the sound, the more orders of diffraction are generated.

As an example, the velocity v of sound in water is about 5000 ft/sec, and at a supersonic frequency f of 150 million/sec (150 MHz), the acoustic wavelength will be given by $\Lambda = v/f = 0.01$ mm, or 10 μm. A light wave entering perpendicular to the cell with a wavelength $\lambda = 0.0005$ mm will be diffracted through an angle θ equal to $\pm 2.86°$ for the first order, $\pm 5.74°$ for the second order, and so on. The velocity of sound in glass is about 11,000 ft/sec, and it is even higher in other materials. Hence, extremely high sonic frequencies are required for acousto-optic devices to be really practical, and it is reported that frequencies up to 3 billion Hz have been generated in the laboratory. Nevertheless, in spite of these difficulties, such devices are becoming commercially available.

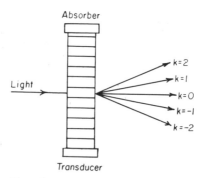

Fig. 16.17. An acousto-optical grating.

Index

A

Abbe, oil-immersion objective, 189
   theory of microscope vision, 196
Aberrations,
   effect on depth of field, 268
   of a lens, 21
Accommodation of the eye, 176
Acoustooptic devices, 314
Adaptation, visual, 93, 170
Afocal Galilean attachments, 223
Afocal systems, 73
Airy disk, 13
   with central obstruction, 260
Aligning an optical system, 5
Altazimuth mounting, 228
Ambient light on a screen, 140
Anaglyph, colored, 140
Anallatic point, 236
Anamorphic systems, 81
   prismatic, 225
Antireflection coatings, 105
Aperture, relative (F-number), 124
Aperture shuttering, in high-speed camera, 291
Apostilb, 119
Apparent brightness of an object, 93, 175
ASA film speed system, 126
Aspheric surfaces, 4
Astigmatism, 22
   visual, 179
Attachments, diopter, 263
Autocollimator, prism testing by, 159
Autofocus mechanisms, 62
Autoset level, 233
AWAR, 61
Axicon, 243
   alignment with an, 244

B

BADOPA plane, 61
Baffles, in mirror systems, 256
Balloon theodolite, 232
Beam section at a tilted mirror, 146
Beam splitters, for color television, 107
   pellicle, 107
Bifocal spectacles, 178
Binocular microscopes, 195
Binoculars, field of view, 205
Biocular magnifier, 185
Birefringent materials, 8
Blaze, of a grating, 311
Borescope, 214
   telecentric relay, 215
Bouwers–Maksutov system, 248
Bravais system, 72
Brewster, polarizing angle, 100
   prism, 224
   teinoscope, 224
Brightness, of object seen by eye, 175
   of telescopic images, 207
   vs. luminance, 176
Bubble sextant, 235

C

Calibrated focal length, 62
Camera, aerial slit, 283
   motion-picture, 286
   optics, 262
   Polaroid, 145
   racetrack, 283
   rotating-mirror, 290

rotating panoramic, 282
rotating-prism, 287
scanning copy, 284
Schmidt, 247
slit, 281
streak, 291
swinging-lens panoramic, 281
telecentric, for orthographic photography, 285
Candela, 94
Candle-power photometers, 111
Cardinal points, 40
Cardioid condenser, 199
Catadioptric systems, 245
 raytracing through, 37
Cataract spectacles, 180
Cat-eye reflector, 246
Catoptric microscope objectives, 192
Catoptric systems, 245
Center of perspective, 262
Central obstruction in mirror systems, 259
Chief ray, 86
CinemaScope system, 81
Cinetheodolite, 231
Circle of confusion, 266
Circular source, 120
Coating, antireflection, 105
 dichroic, 107
Coelostat, 151
Collimator, 55
Coma, 21
 correction of, in a telescope, 256
 of elliptical mirror, 132, 249
 of parabolic mirror, 250
Communication fibers, 24
Condenser, in slide projector, 133
Conjugate distances, of a thin lens, 53
 relation between, 47
Constant deviation spectroscope, 304
Constant deviation system of two mirrors, 153
Contour projector, 141
Coronagraph, 228
Cos^4 law, 122
Critical angle, 11, 99
Critical flicker frequency, 174
Crystals, optical, 3
Curvature of spectrum lines, 299
Curved mirror, raytracing at, 36
Curved screen, 138
Cylinders, crossed, 83
Cylindrical surfaces, 5

D

Daguerreotypes, image inversion in, 146
Dark field illumination, 199
Dawes' rule, 207
Densitometers, 109
Density, 108
Depth of field, 265
Diapoint, 26
Diffuse density, 108
Diffusing surface, 93
DIN film-speed system, 127
Diopter, 45, 176
 attachments, 263
Dispersion, of a grating, 310
 of a prism, 296
Displacement of image by a parallel plate, longitudinal, 157
 transverse, 241, 288
Distortion, anamorphic, 225
 when rectifying keystone, 220
 with a cylindrical Bravais system, 73
 curvilinear, 21
 effect on oblique illumination, 129
 in a panoramic camera, 282
 keystone, 269
 vanishing point, 269
 of a telescope eyepiece, 205
 wide-angle, 262
Double monochromator, 304
Dynameter, 204
Dyson, interference microscope, 200
 relay for a microscope objective, 245

E

Ebert–Fastie spectograph, 312
Edge gradient, 20
Ellipse, drawing an, 147
Endoscopes, 226
Entrance pupil, 86
 of fish-eye lens, 87
Equatorial mounting, 228
Equivalent refracting locus, 40
Erecting telescopes, 210
Errors, in mounting a rotator, 163
 in prism angles, 159
Exit pupil, 86
 of a microscope, 190

INDEX

of a telescope, 210
spherical aberration of, 210
Exposure equation, 126
Exposure meters, 117
Extension tubes, 263
Eye, accommodation, 176
  adaptation, 170
  as a photometer, 110
  as an optical instrument, 170
  dark adaptation of, 170
  dimensions of, 170
  far point, 176
  near point, 176
  pupil diameter, 171
  resolving power, 172
  spectral sensitivity, 174
  stereo acuity of, 180
  wavelength sensitivity, 174
Eye-level viewfinder, 222
Eyepieces, 208
  Delaborne, 208
  Erfle, 208
  four-lens terrestrial, 210
  Huygens', 208
  Kellner, 208
  military, 210
  Ramsden, 208
Eyepoint, of a microscope, 190
  of a telescope, 204
Eye relief, 210

F

'F-Θ' lens, 153, 274
Fiber optics, 22
  communication channel, 24
  endoscopes, 226
  face-plates, 23
  image invertor, 23
  wave-guides, 24
Fiberscope, 23
Field curvature, 22
Field, depth of, 265
Field flattener, see Lens, Smyth
Film projector, 8mm, 132
Flare spot, 128
Flicker, 174
Flux, 95
  measurement of, 112
  radiated into a cone, 117
Flux meters, 112
Flying spot scanner, 292
F-number, effective, 84
  of a lens, 124
Focal lengths, 40
  calibrated, 62
  measurement of, 59
  relation between, 42
Focal points, 40
Foco-collimator, 61
Focometry, 59
Focus, depth of, 265
Focusing a camera on a near object, 263
Foot-candle, 96
Foote's formula for oblique illumination, 123
Fourier transform plane, 202
Fractions of a stop, 125
Frequency of light waves, 7
Fresnel, formulas for reflection, 98

G

Galilean telescope, 221
Galilean viewfinder, 222
Gauss theory of lenses, 40
Ghost images, 128
Glass, optical, 2
Gradient index, materials, 3
  GRIN rods, 24
Graphical image construction with thin lenses, 66
Grating, diffraction, 306
  blaze, 311
  concave, 312
  dispersion, 310
  magnification, 308
  resolving power, 310
GRIN rods, 3
Gullstrand ophthalmoscope, 214

H

Hadley's sextant, 234
Head-up display, 252
Heliostat, 151

318   INDEX

Hiatus, 52
High-speed cameras, 286
Homogeneous immersion, 189
Hot spot, with translucent screen, 139
'Hycam' camera, 289
Hyperopia (far sight), 177

Intensity, 94
  measurement of, 111
Internal reflection, 11
Interpolation of refractive indices, 12
Inverse square law, 95
Iris and pupils, 86
  of a telescope, 204

I

K

Illuminance, 95
  from a circular source, 120
  measurement of, 112
Illuminance meters, 112
Illumination, Abbe, for a microscope, 197
  dark field, 199
  in an optical image, 124
  Köhler, for a microscope, 198
  for a slide projector, 133
Illuminator, vertical, 190
Image, contrast, 19
  displacement by a parallel plate, 157
  in a large field lens, 185
  left-handed, 144
  real, 9
  right-reading, 144
  virtual, 9
  stray light in, 128
  wrong-reading, 144
Image brightness, to the eye, 175
  in a telescope, 207
Image contrast, 19
Image erector, prismatic, 164
Image illuminance, distant object, 124
  near object, 125
Image inversion, in a slide projector, 144
Image invertor, fiber optic, 23
  prismatic, 164
Image lean, in prism binocular, 167
Image magnification, in afocal system, 205
Image processing, 202
Image rotators, 161
  mounting of, 163
Image shift, methods for producing, 158
Image space, 10
Image tracking through lenses and mirrors, 160
Immersion objectives, 189
Infrared, materials, 3
  telescope, 227
Integrating sphere, 114

Keystone distortion, 269
  in overhead projector, 135
  rectification of, 270

L

Lagrange theorem, distant object, 43
  near object, 37
Lambert, cosine law of illumination, 96
  cosine law of intensity, 94
  unit of radiated flux, 119
Lambertian diffuser, 93
Laser beam, expander, 222, 255
  used for alignment, 5, 244
Laser rangefinder, 237
Lateral color, 21
  of eyepieces, 208
Left- and right-handed images, 144
Lens, anallatic, 237
  anastigmat, 22
  ball, see Sutton's panoramic
  centering and edging, 4
  contact, 180
  cylindrical, 80
    crossed, 83
    in Bravais system, 73
    used to make a sine-wave bar chart, 82
  diamond grinding of, 3
  diffraction limited, 16
  erecting, 210
    in telescope, 210
  field, at translucent screen, 139
    image projection into, 186
    in a telescope, 213
    in orthographic photography, 285
    of an eyepiece, 209
  fish-eye, 274
    entrance pupil of, 87

Fresnel, 38
liquid-filled, 3
monocentric, 275
photographic, 280
plastic, aspheric, 5
reversed telephoto, 71, 274
Ross, for coma correction, 257
sky, *see* fish-eye
spectacle, 176
Smyth, 213
surface power, 45
Sutton's panoramic, 275
telecentric, in contour projectors, 141
telephoto, 71, 272
thick single, 51
thin, 52
    raytracing through separated, 67
toric, 179
zoom, 277
    mechanically compensated, 277
    optically compensated, 279
    two-component, 74, 278
Lens erector in a telescope, 210
Lens manufacture, 3
Lens power, 45
Lensometer, 49
Level, autoset, 233
    surveyor's 232
Light and color, 92
    detectors of, 7
    nature of, 7
    polarized, 8
    sources of, 7
    wavelength, 7
Light adaptation of eye, 170
Littrow spectrograph, 302
Longitudinal magnification, 56
Lumen, 95
Luminance, 93
    measurement of, 112
    of an aerial image, 97
    of an illuminated surface, 119
    vs. brightness, 176
Luminance meters, 112
Lux, 95
Lyot stop, in coronagraph, 229

M

Macro lens, 265

Macro zoom, 265
Magnification, 36
    in a telescope, 205
    in an afocal system, 74
    longitudinal, 56
    of a grating, 308
    of a prism, 297
Magnifying instruments, 182
Magnifying power, 182
    of a microscope, 187
    of a telescope, 203
Magnitude of stars, 176
Maksutov–Bouwers systems, 248
Maksutov Cassegrain system, 256
Materials, optical, 2
Matrix theory of paraxial rays, 77
Maxwellian view, 97
Mean spherical candle power, 94
Meridional rays, 26
    tracing of, 28
Meter-candle, 96
Microdensitometer, 110
Microphotography, 194
Microscope, 187
    binocular, 195
    compound, 187
    interference, 200
    spatial filtering in, 197
    tube length, 187
    Zernike phase, 201
    zoom, 192
Microscope objective, numerical aperture of, 188
    oil immersion, 189
    Schwarzschild, 258
Microscopy, of transparent objects, 199
    ultraviolet, 189
Military rangefinder, 238
Mirror, beam section on a tilted, 146
    conic-section, 248
    elliptical, 248
    hyperbolic, 253
    Mangin, 247
    off-axis paraboloid, 250
    parabolic, 250
    plane, flatness tolerance of, 147
        flexure, 147
    triple, 154
    rotating, 148
    spherical, 245
Mirror drum for scanning, 152

Mirror imaging systems, 245
Mirror microscope objectives, 258
Mirror scanners, 152
Mirror systems, central obstruction in, 259
Modulation transfer function, 17
  cascading, 19
  coherent, 20
  with central obstruction, 261
Monochromator, 302
  Littrow, 302
  van Cittert double, 305
  Wadsworth, 252, 303
Mounting an image rotator, 163
MTF, see Modulation transfer function
Multifocal spectacles, 178
Multi-lens systems, 66
Multiple mirror systems, 153
Myopia (near sight), 177

N

Near object, focusing on a, 263
Nearest distance of distinct vision, 182
Night-glass effect, 173
Nit, 94
Nodal points, 50
  of a mirror, 246
Nodal slide, 59
Numerical aperture, 17
  of a fiber, 22
  of a microscope objective, 188

O

Object space, 10
Objects, real, 9
  virtual, 9
Oblique beams, 85
  paraxial tracing of, 89
Oblique image illumination, 129
Obstruction, in mirror systems, 259
Oil-immersion microscope objective, 189
OPD, see Optical path difference
Ophthalmoscope, Gullstrand, 214
Optical fibers, 22
Optical Invariant, 38
Optical path difference, 13

Optical square, 234
Optical system components, 2
  alignment, 5
  design and production, 1
Optical tunnel, 156
Orthographic photography, 285
Overhead projector, 135

P

Panoramic camera, 281
Paraboloid condenser, dark field, 199
Parallel beam, 55
Parallel plate, image displacement by,
  longitudinal, 157
  transverse, 241, 288
Paraxial rays, 26
  matrix theory of, 77
  tracing, 30
Perfect optical system, 13
  resolving power of, 16
Periscopes, 211
  stereoscopic, 181
Perspective, in photography, 262
Phase microscopy, 201
Phot, 95
Photographic exposure meters, 117
Photographic lens, specifying a, 280
Photographic optics, 262
Photography, underwater, 275
Photometric definitions, 93
Photometric measuring instruments, 110
Photometry of optical systems, 92
Photomicrography, 194
Photostat machine, 145
Pile of plates, 105
Plane grating spectrograph, 312
Plane surfaces, photometry of, 98
Plane-table surveying, 230
Polarization by reflection, 100
Polarized light, 8
Polaroid cameras, image inversion in, 145
Polaroid sheet material, 8
Polygon mirror scanners, 152
Polygon, reflecting, 152
  transmitting, 287
Power changers, stepped, 217
Power of a lens, 45

INDEX

Principal planes, 40
  relation between, 41
Principal points, 40
Principal ray, 86
Prism angle errors, 159
Prism autocollimator, 159
Prism binoculars, field of view of, 205
  used as a magnifier, 206
Prism, Brewster, 224
  double-image, 102
  Nicol, 102
Prism monochromator, 302
Prism spectrograph, 300
Prismatic image erectors, 164
Prisms, dispersing, 294
  curvature of spectrum lines, 299
  deviation, 294
  dispersion, 296
  magnification, 297
  minimum deviation, 295
  Pellin–Broca, 303
  resolving power, 298
Prisms, reflecting, 155
  Abbe, 161
  Amici roof, 155
  cube-corner, 154
  Daubresse, 167
  Dove, 161
    rotating object seen through, 163
  erecting, 164
  for image rotation, 161
  Hensoldt, 166
  inclined eyepiece, 155
  Leman, 166
  Möller, 166
  optical tunnel of, 156
  Pechan, 161
  penta, 155
  Porro, 165
  right-angled, 155
  roof, 155
  Schmidt, 161
  Taylor, 161
  Uppendahl, 161
  Wirth, 165
Projection, stereoscopic, 140
  three-dimensional, 140
  with illuminated object, 131
  with self-luminous object, 131
  zoom lenses for, 279
Projection screens, 138

Projector, arc, 131
  contour, 141
  Eidophor, 137
  Kodak contour, 142
  slide, 133
    image inversion in, 144
Pupil diameter of eye, 171
Pupils, entrance and exit, 86
  of a telescope, 204
Purkinje effect, 175
Pyramid error of a prism, 159

Q

Quartz, spectrograph, 301
Quasi-parallel beam, 56
'Questar', optical system of, 256

R

Radiometry and photometry, 92
Rangefinders, 236
  laser, 237
  military, 238
  self-contained-base, 238
  stadiametric, 236
  stereoscopic, 242
Rayleigh limit of optical path difference, 13
Rays and waves, 8
  types of, 26
Ray tracing, 27
  at a curved mirror, 36
  graphical, 27
  meridional, 28
  paraxial, 30
  through separated thin lenses, 67
Real and virtual images, 9
Rectification of keystone distortion, 270
Reflaxicon, 244
Reflection, at a glass surface, 98
  at normal incidence, 103
  total internal, 11, 98
Refraction, law of, 10
Refractive index, 7
  interpolation of, 12
Resolving power, eye, 172
  grating, 310

microscope, 188
perfect lens, 16
prism, 298
telescope, 206
Rifle scope, 215
Rim rays, upper and lower, 85
Roof edge, precision needed in, 159
Rotators, 161
   mounting a, 163
Rowland circle, 312

S

Scanners, mirror, 152
Scheimpflug condition, 58
   in a rectifying enlarger, 270
Schlieren system, 251
Schwarzschild microscope objective, 258
Screen, curved, 138
   diffusing, 139
   opaque, 138
   projection, 138
   translucent, 139
Screen gain, 120, 138
Screen lumens, measurement of, 137
Selfoc GRIN rods, 24
Sextant, 234
Shifting an image along axis, 158
Siderostat, 149
Sine-wave bar chart,
   made by cylindrical lens, 82
   used for MTF measurement, 17
Skew rays, 26
Slide viewer, biocular, 185
Snell's law, 11
Spectacle lenses, 176
Spectrograph, grating, 312
   prism, 300
Spectrophotometers, 116
Spectroscopic apparatus, 294
Specular density, 108
Stadiametric rangefinder, 236
Stellar magnitudes, 176
Stepped power changer, 217
Steradian, 95
Stereoscopic, aerial slit camera, 283
   binoculars, 167
   microscope, 195
   periscope, 181

   photographs, 181
   projection, 140
   rangefinder, 242
   viewing in a field lens, 187
   vision, 180
Stigmatic image, 22
Stilb, 93
Stray light in images, 128
Surface reflection, 98
Surveying instruments, 230
Surveyor's level, 232
Swinging collimator, for focometry, 61
Symmetrical systems, 21

T

Teinoscope, 224
Telecentric systems, 88
Tele-extender, 265
Tele-microscope, 227
Telenegative attachment, 265
Telephoto system at finite magnification, 69
Telephotometers, photoelectric, 113
   visual, 113
Telescope, 203
   astronomical, 227
   Cassegrain, 254
   coma correction in, 256
   Dall–Kirkham, 255
   declination axis of, 228
   Donders, 223
   erecting, 210
   field stop, 204
   fields, real and apparent, 205
   Galilean, 221
      as a power changer, 219
      mounted in hotel room door, 222
      used in front of camera lens, 223
      used in submarine periscope, 212
   Gregorian, 254
   image brightness in, 207
   infrared, 227
   McMath Solar, 150
   Newtonian, 254
   pancratic, 211
   panoramic, 168
   polar axis of, 228
   resolving power of, 206
   Ritchey-Chrétien, 257

siderostatic, 150
space, 207
unit-power, 213
with stepped power changer, 218
zoom, 211
Theodolite, 231
Thickness, insertion of, 54
Through-the-lens metering, 117
Tilted planes, 58
  theory of, 269
Tilted projection, 271
Total internal reflection, 11, 98
Tracking an image through lenses and mirrors, 160
Transit, surveyor's, 231
Transmission at normal incidence, 103
Transmittance measurement, 115
Transparent objects under a microscope, 199
Trifocal spectacles, 178
T-stop system, 127
Tube length of microscope, 187
Tungsten iodine lamps, 134
Two-mirror systems, 153, 253

U

'Ultrapak' illuminator, 191

Unit planes, 42
U.S. aperture designation, 127

V

Veiling glare, 128
Vertometer, 49
Viewfinder, Galilean, 222
Vignetting, 85
Virtual objects and images, 9
Vision, properties of, 172
Visual brightness, 175
Vuegraph, 135

W

Wave number, 7
Waveguides, fiber-optic, 24
Waves and rays, 8
Waxicon, 244

Y

Y-nu method for tracing paraxial rays, 34